D0742009

MASTERPIECES OF RUSSIAN FOLK ART

V. BORISOVA

Rostov Enamels

INTERBOOK · MOSCOW · 1995

Text by:
Valentina Borisova

Photo by:
Anatoli Firsov

Illustrations compiled by:
Valentina Borisova, the State Russian Museum
Svetlana Guseva, Rostov Finift Factory
Marina Fedorova, Rostov-Yaroslavsky
Museum Preserve of Architecture and Arts

Layout and book design by:
Juri Leonov in collaboration with **Vassily Kiselyov** and **Anatoli Savyolov**

Translated by:
Liudmila Lezhneva

ISBN 5-7664-1056-5

... Enamel
is a type
of glass
imbued
with paint.

V. K. Trediakovsky

The mention of Rostov enamels brings to mind precious miniatures which have for centuries adorned church utensils, household objects and pieces of jewelry. Bright colours glowing against white enamel plaques brought to life scenes of biblical and Russian history and mirrored multifarious life and the beauty of native land. Masterful craftsmen of Rostov the Great, in which a unique technique of enamelling was carried on for over two centuries, are producing these treasures to our day. In the past, craftsmen of the Rostov bishopric and the monasteries painted, on commission from the clergy, enamel plaques to decorate icons, chalices, reliquaries and other cult objects. Later on townsfolk took up the expensive craft of enamelling, passing its secrets from father to son, from one generation to another. Less dependent on the clergy and guided by client interests and their own tastes, the artisans extended the boundaries of enamel miniature genres, adding to them portraits and plaques showing the Rostov Kremlin and monasteries.

The decorative nature of miniature paintings, which could embellish both big and small objects, ensured the longevity and the wide application of Rostov enamels. They shine with motley hues in jewelry pieces produced by today's craftsmen. A frame of metal wire twining in fancy patterns complements exquisite enamel miniatures. The painters and jewellers pool efforts to create extraordinary integral works of art, be it a decorative panel, a casket, a portrait or women's jewelry. Though not indispensible, these things add colour to our households and life, giving joy and warmth to the inhabitants of this cold industrialised world. When we look at these miniature pictures of Russian nature or old history, we are brought back to our sources and eternal values.

In its two-century history Rostov enamels have repeatedly gone through the periods of florescence and decline. The main theme of the craft — a tribute to the beauty of the native land and to man's lofty spirit — was, however, never abandoned. Rostov enamellers have preserved to our day and carried on the best traditions of Russian enamel miniature painting. The craft never existed in isolation: while developing the local tradition of icon painting, it came under the influence of other schools of painting. That is why we begin the story of Rostov enamels with the history of enamelling in Russia.

Enamel is one of the oldest forms of decorating metal. It is a vitreous substance, to which metal oxides give different colour. Since very early times enamel was used to decorate objects made of gold, silver and copper. The process was known in Ancient Egypt and through Byzantium it was transmitted to Russia.

The first Russian mention of enamel goes back to the reign of the Duke Andrei Bogolyubsky of Suzdal. The Ipatievskaya Chronicle (1174—1175) says that the Church of the Nativity of the Virgin built on the Duke's Grand Order was decorated with «gold, enamel and all type of virtue.»[1]

The technique of cloisonné enamel was known to have been employed in Kievan Rus way back in the 11th century. It was referred to as o finift,п which is the Old Greek for alloy or shining stone. The vitreous surface of enamel, decorating old ritual vessels, women's diadems, *kolt*-pendants, *barmy*-shoulder pieces and so on, indeed shone with a deep light of precious stones. Cloisonné enamel went perfectly well with precious stones that Russian craftsmen used to decorate their wares. Born in a glowing kiln, the opaque mass of special glass, mixed with metallic oxides, filled the surface of a picture picked out by thin partitions on a gold plaque. The enamel melted in the heat, acquiring mellow colours and fusing with the metallic base. The shining gold background and the shimmering outlines perfectly harmonized with the exquisite colour scheme of cloisonné enamel.

The decorative potentialities of enamel attracted gold- and silversmiths at all the times. For centuries on end they polished the technique of enamelling and various

[1] Ipatievskaya Chronicle (1174—1175), Vol. 2. Archaeography Commission Publication, St. Petersburg, 1843, p. 111.

processes. Jewelers mastered the process of covering a cast and incised picture, an engraved ornament or a design outlined by thin metal bands with the enamel paste. It was not, however, until the 17th century that boundless prospects opened for enamelling when the French craftsman Jean Toutin discovered the secret of painted enamels. To make an enamel miniature, jewelers painstakingly embossed gold, silver or copper plaques and then applied glass and opaque enamel layer by layer. After repeated application of enamels and firings the plaque acquired a smooth and glossy surface, onto which the design was then applied in colours with different melting temperatures. To attain glowing colours and diverse shades, enamellers would fire an enamelled plaque up to half a dozen times and often resorted to techniques employed by painters and graphic artists. They shaped the form with thick strokes, using the white background for the openings. To make the picture three-dimensional and to convey air space, painters used dots and strokes of different size, thickening in the shade and loosening in the light. Enamel miniatures retained their traditional decorative functions, meanwhile the vast opportunities offered by the media allowed enamellers to extend its boundaries. They worked hard on enamel portraits and landscapes and created multi-figural religious, historical and genre compositions.

Russian craftsmen had mastered the complicated technique of enamel painting by the late 17th century. Artisans working at the Armoury Chamber and the northern town of Solvychegodsk produced various objects, including caskets, goblets and flasks, decorated with enamel. The design basically consisted of large ornamental motifs with twining shoots and garlands of gorgeous flowers — tulips, corn-flowers and sun-flowers. The folk craftsmen's imagination and inventiveness blended foliate ornaments, fantastic motifs and pictures of the surrounding world into a single decorative composition. Referred to as Usolie design, that type of painting, for all its conventionality, conveyed the Russian enamellers' interest in the beauty of their natural environment. Later on that interest became manifest in the development of portraiture, landscape painting, still lifes and other genres.

As time went by enamels reflected the peculiarity of every historical period, the changing general styles of art and the desires and tastes of the customers.

For example, the appearance of enamel portraits in Russia is usually associated with the reign of Peter the Great who admired extraordinary and gifted personalities. The wide spread of caskets, snuff-boxes and boxes bedecked with portraits of royalties and other eminent citizens is explained by the passion of Russian empresses for all sorts of gifts and expensive bric-a-brac.

In harmony with the spirit of the times, two parallel trends — decorative painting and easel miniature — developed in 18th-century Russian enamels. These two trends often coincided in some enamels decorating household and cult objects, imparting a meaningful content to a purely decorative form. These qualities were especially graphically manifest in miniatures on religious themes, in Kiev works in particular. Kiev enamels, with their highly decorative colour scheme, were very expressive. The beautiful images of saints at times though seem inordinatedly exhalted. The best enamellers were especially skilled at conveying different psychological states and human characters. Their pictorial idiom matched the inner energy of images and plots. Miniatures depicting dramatic events from the life of Christ were often painted against a black or stormy background, with intense colours and dynamic compositions sustaining the emotional fervour of the plot.

Until the late 18th century Ukrainian religious enamel miniatures retained the baroque stylistic idiom. Like other types of applied arts, however, Russian enamellers were sensitive to any change in the general artistic style. While Moscow enamellers drew on the local artistic traditions for quite a while, their St. Petersburg counterparts promptly embraced the accomplishments of European enamellers. Foreign jewelers visiting St. Petersburg had a strong influence on the development of St. Petersburg's miniature style. In the second half of the 18th century the luxurious baroque style gave way to austere Classicist forms. The new artistic ideas were promulgated by the St. Petersburg Academy of Arts, where a class of miniature painting opened in 1779. That fact bespoke of the official recognition of miniature which became quite popular with every household.

Order badges, which were mostly made in Moscow and St. Petersburg, formed a special branch of Russian enamelling. On one occasion, however, when the archbishop Arseny Vereshchagin of Yaroslavl was awarded the Order of Alexander Nevsky, he commissioned two enamel plaques with the representations of the saint to be made for the Order cross. They were painted by a o theological sextonn , who could be a resident of Rostov[2]. The enameller copied the original, which had been sent to the archbishop from St. Petersburg. Provincial craftsmen thought highly of works of art produced in the Russian capital. They imitated and copied pieces produced

[2] Arseny Vereshchagin's Diary (1786—1799), Vol. 2. Entry of August 14, 1797. Russian National Library, Manuscript Department, 0 IV, No 267

by painters trained at the Academy of Arts. All that notwithstanding, Rostov enamellers felt closer affinity with their Moscow colleagues.

Enamelling did not start from scratch in Rostov, whose icon painters and silversmiths were well acclaimed in the past. The Rostov bishopric was known to have had an icon painting workshop since olden days. It is hard to say when the expensive craft of enamelling struck root in that provincial town. Legend has it that an exiled Italian enameller taught his craft to local icon painters in the reign of Empress Anna Ioannovna, but there is no documentary evidence of that fact.

Many scholars sought to shed light on the obscure history of the emergence of Rostov enamels. Some connected it with the activity of metropolitan Iona Sysoyevich, who contributed to the development of the Rostov bishopric in the late 17th century. Others cited the rule of archbishop Ioakim, who founded a Graeco-Latin school there on Empress Anna Ioannovna's order. It is to that period that the legend of the exiled Italian enameller dates back. No evidence confirming either theory has been found, meanwhile the rare mention of objects decorated with enamel in late 17th and early 18th century documents gives little ground to suggest that enamelling had already emerged as a local business. Anyhow, the earliest mention of an enamel workshop in the inventories of the Rostov bishopric appeared in the 1760s. It was perhaps at that time that, following a synodic order, the Rostov metropolitan's province and the monasteries began to attract gifted people to be trained not only in icon painting but also enamelling. Craftsmen from among the clergy and assigned peasants executed exclusively church orders. Inventories of the bishopric warehouses give an idea of the range of works they performed. There were individual enamel plaques for icons depicting Rostov and other Russian saints held in especially great esteem, as well as miniature sets for church utensils, metropolitan's hats and Gospel frames. Some of them were sent to Yaroslavl, the fact which prompts the conclusion that in the 1760s the Rostov enamel workshop served clients well beyond the town's boundaries[3]. Invaluable documents of that period preserved the name of enameller Gavrila Yelshin, who served at the Rostov bishopric. We know nothing of his life and work, nor have his works, quite possibly extinct by now, been unearthed. Gleaned carefully in archival documents, the available data allow to imagine the colour scheme of his works. Among the paints which were to be issued to Yelshin were ordinary purple of ground gold, rouge and violet; corporeal, thick and ordinary yellow; three shades

[3] *Fedorova M. M.*, On Early Rostov Enamels. Rostov Museum Reports, Rostov, 1993, p. 128.

of green; the tausin colour that is unknown to us — a total of 19 colours. They formed the rich baroque palette characteristic of many works of Rostov enamellers of the late 18th — early 19th centuries[4].

The bishopric workshop was operational until the late 1780s, fulfilling clerical orders from different Russian cities and towns. After the metropolitan's office was transferred to Yaroslavl, some enamellers continued to execute the bishopric's orders, while other, such as the Isayevs brothers were granted relative independence and came to be registered at the enamel workshop of the Rostov artisan tribunal. That fact can be considered as a starting point in the development of enamelling as a full-fledged trade in the town. The Rostov bishopric workshop had an important role to play in the evolution of Rostov enamels.

The Reverend Alexei Ignatievich Vsesvyatsky (1762-1831), a celebrated local enameller, could have been trained at that workshop. Vsesvyatsky came from a family of clerics and icon painters and fulfilled the most important orders, painting enamel miniatures for utensils for the Cathedral of Assumption and other Rostov churches[5]. On the order of the first Yaroslavl archbishop Arseny Vereshchagin Vsesvyatsky made, in 1791, a plaque with Platon Levshin's monogram to be presented to the Moscow metropolitan in memory of his visit to Rostov the Great[6]. Vsesvyatsky had a perfect command of painted enamels techniques. His style and interpretation of faces and other uncovered parts of the bodies, as well as that of the remaining background landscape, architecture, utensils and clothes betrayed the influence of the old icon painting tradition. It was only natural that his experience of an icon painter left an imprint on the stylistics of his works.

An enamel mount painted for the altar cross for the Cathedral of Assumption reproduces the iconographic pattern that was typical of Russian religious painting starting from the 17th century. Vsesvyatsky portrayed the key Gospel scenes of the *Resurrection* and the *Descent into Hell* on an oval plaque one above the other. The central figures of the composition are enframed with a band, shaped as the number eight and bearing an inscription. The shape of the scroll and the inscription on it, which gives a peculiar interpretation of the biblical parable, remind of the eternal flow and renovation of life, imbuing the miniature with a profound philosophical meaning. That ring of eternity seems to separate the central figures from the vain and perishable world. The bright halo around Christ alone picks out the silhouettes

[4] Ibid., pp. 127—128.

[5] *Zyakin V. V.*, New Documents on the Life and Work of the Rostov Enameller A. I. Vsesvyatsky. Year-Book: Cultural Monuments. New Discoveries, Moscow, 1988, p. 355.

[6] Arseny Vereshchagin's Diary (1786—1799), Vol. 2. Entry of January 31, 1791.

from the void, creating an impression of a phantasmagoric space. The contrast of light and dark conveys the dramatism of what is going on, which is further enhanced by the intense colour scheme. The latter simultaneously increases the overall decorative effect. Following the accepted canon, the enameller did not forget about the applied nature of his work. He used bright local colours for clothes, resorting to whitening and highlighting as icon painters usually did in hatching. The central figures were sharply outlined, if only schematically in keeping with the old instructions. The faces of the saints, in accordance with the icon painting tradition, were magnanimous and benignant, while at the same time retaining national characteristics and mirroring the best Russian features, such as kindness, simple-heartedness and artlessness. Showing a masterful command of enamelling techniques and pictorial effects, Vsesvyatsky remained a truly folk artist, who developed the local tradition in his works. There was something in common between his works and many Moscow enamels which bore an imprint of the Armoury Chamber style until the early 19th century. They shared a similar understanding of the icon painting tradition, the baroque nature of painting and a certain, slightly uncouth type of popular characters.

The Rostov enamel tradition is, however, far more versatile. It developed for centuries in close contact with other major art schools and grand styles. From the outset Rostov enamellers espoused the flamboyant symbol-laden baroque style and evolved, on its basis, their own artistic tradition.

Craftsmen whose work was connected with the Spaso-Yakovlevsky Monastery played an important role in the evolution of Rostov style. Founded in the late 14th century, the monastery became, starting from 1757, a place of worship of the relics of a new Russian wonder-worker, Metropolitan Dmitri of Rostov. Though it is not known whether the monastery had at that time an icon painting or an enamelling workshop, documents and legends confirm that in the last quarter of the 18th and the first half of the 19th century artists from among the service people and the monks painted icons and holy images in enamels to be presented as gifts to guests and rich donators[7]. Those might have been the craftsmen who also produced enamel plaques that the monastery needed to make ritual objects, such as mitres, chalices, Gospels and altar and pectoral crosses. Some of these objects, which are noteworthy for their rich and exquisite decor, have survived at the Rostov-Yaroslavsky Museum Preserve. Church utensils and crosses were decorated with precious metals and gems; the mitres

[7] Rostov-Yaroslavsky Museum Preserve of Architecture and Arts, Manuscript Department, unit No R-887, sheets 2, 3, 7.

were embroidered with gold and silver thread and pearls and studded with gems and paste. Mounts with painted enamels, however, occupied pride of place.

Some miniatures are identical in style and the painting technique employed, as if they were done by one and the same craftsman. Of special interest is an enamel plaque with the scene of *Crucifixion*, which was painted for an altar cross commissioned by Archimandrite Amphilochy and which was kept as a sacred object in the sacristy. The reverse counter-enamel bears the initials о А Мп and the manufacture date «1777.»

The name of Alexander Grigorievich Moshchansky, whose initials coincide with the above signature often crops up in the documents of the Spaso-Yakovlevsky Monastery. Moshchansky (1745—1824) lived in the Yakovlevsky suburb outside the monastery and was registered on the monastery staff[8] until 1814. He painted on commission portraits of St. Dmitri of Rostov and scenes from religious history and made enamel plaques for ritual objects and enamel images of saints.

Enamel plaques which show SS. John the Baptist and Dmitri of Rostov and bear the signature of Alexander Moshchansky, can be found in the collections of the State History Museum and the Hermitage. These pieces, with their colourful painting and dynamic composition, are close in style to the *Crucifixion* miniature. The similar colour scheme, in which violet, purple, green and brown predominate, is produced of intermediate and complementary colours. The painting techniques used in these enamels are also identical: now fading, now thickening, the strokes convey the play of light and shade, with the halos formed by whirling strokes. The foldsof the draperies, in little conformity with the shape, bend sharply and break with intensity, subordinate to the inner rhythm of the decorative composition. The images of the saints are somewhat schematic, though lit up with the flames of human emotions. The choice of the composition and pictorial solution betrays the influence of West European engraving and of Ukrainian models, the latter apparently not accidental. Let it be recalled that the monastery was actively rebuilt under the archimandrites of Ukrainian descent, Amphilochy Leontovich and Avraamy Florinsky. Even before that, in 1762, Venedict Dmitrievich Vendersky, an artist invited from Kharkov, worked at the monastery[9].

Either he or other Ukrainian artists could have influenced the evolution of a special pictorial style of Moshchansky and other craftsmen of his circle. The icon painters

[8] Central State Archives of Old Acts, fond 1407, inv. 1, units Nos 1096, 1159.

[9] Rostov Branch of the State Archive of the Yaroslavl Region, fond 145, inv. 1, unit No 11. Book of Historical Records for 1836, the 1762 entry. According to other data, it could have been another Kharkov painter, Venedict Dmitrievich Svidersky.

and enamellers working at the Spaso-Yakovlevsky Monastery retained the lofty spirit and emotional idiom of the Ukrainian and Russian baroque style until the mid-19th century.

The pictorial style of Rostov enamels began to be renovated in the first half of the 19th century both due to a change in taste and due to the impact of the classicist principles of Rostov's temple architecture and its interior decoration. Enamel miniatures which decorated church utensils formed but a minute segment of the entire ensemble and, naturally, adopted the pictorial idiom of the new style. Rostov enamellers used as models engravings from paintings by West European and Russian artists, as well as numerous original religious paintings that landed in Rostov monasteries and churches in the form of donations.

Rostov craftsmen sold their wares in different Russian cities and towns and, when visiting the capital, could see works by Academy artists. Less tied up with religious canons, enamellers living in cities were faster to assimilate classicist pictorial techniques. They naturally took in iconography and traditional shapes, as well as the lofty aesthetic ideals of the new style. Enamellers regarded themselves as artists rather than craftsmen and sought to attain the heights of St. Petersburg's Academy painting. It was not by chance that the best masters, such as Nikolai Salnikov and Roman Vinogradov, presented their works at the Academy of Fine Arts, seeking the title of an artist without rank or station.

By the mid-19th century there were about 50 enamellers in Rostov, some of them running their own businesses but the majority working at home. United by a trade corporation, the enamellers remained independent both in their work and in marketing their products. The best of them retained individuality and their own original idiom. The Shaposhnikovs, Zavyalovs and Rykunins families of enamellers were known well beyond their native town. Works of Yakov Ivanovich Shaposhnikov, who came from a family of enamellers, were distinguished by superb craftsmanship and intricate designs. He was the trade corporation's foreman, who took the most important commissions. His enamel images of saints and miniatures for large hagiographic icons have masterful compositions and harmonious colour schemes. Shaposhnikov's style was romantic, and the images he painted were sublime. Despite a certain stereotype quality, they seem compassionate, understanding and, as it were, conveying the maxim formulated by Alexander Pisarev, an exponent of classicist ideas, «The arts must without fail preach morality, edify and grace the popular spirit so as to both comfort and instruct.[10]»

The special atmosphere of a provincial town and the affinity between the local townsfolk and peasants could not but influence the development of enamelling. Though guided in general by St. Petersburg trends, Rostov enamelling remained true to the values of local folk culture in the first half of the 19th century. The works of Alexei Gavrilovich Tarasov and Yakov Ivanovich Rykunin upheld the common Russian pictorial tradition and local art.

Tarasov's enamel icons are characterised by especially gentle technique and images. The faces of his saints are lit up by a divine glow, which makes the robes shine even brighter.

Rykunin felt less restrained in the choice and interpretation of iconography. He followed both the icon painting tradition and classicist models, when painting enamel miniatures for *The Resurrection of Christ with Scenes from His Life* (1854, SRM)[11]. An enamel plaque showing *Resurrection and Descent into Hell* is central to the icon. Like Vsesvyatsky, Rykunin reverted to the old Russian tradition by placing the main Gospel scenes successively one above another along the surface of the plaque. He created a special picturesque world, conveying the infinite beauty of the environment with the help of vigorous translucent colours. Rykunin showed less freedom, however, when copying engravings from Academy artists' paintings. It was only in *The Flagellation of Christ* done after an engraving from Yegorov's painting that Rykunin slightly departed from his model by showing a mocking soldier to the right from Christ. His interpretation revealed his interest in the Bible text and various mundane details. Popular world outlook, high professionalism and uncommon pictorial interpretations naturally combined in his work. Together with other leading painters of the mid-19th century, Rykunin sought new ways of developing Rostov enamels.

There also were other craftsmen, who copied the original models, multiplied, modified and polished new iconographic motifs. Now striving after a terse solution, now guided by their own taste, enamellers changed compositions, discarded superficial details and varied the colour scheme. For example, both remarkable painters and ordinary artisans worked together with Ivan Zavyalov. His workshop turned out both sumptuous icons picked out in gold and small plain images, all bearing the same trademark on the frame. The earlier products complied with the classical canons

[10] *Pisarev A., Objects for Artists,* Part 1, St. Petersburg, 1807, p. 6.

[11] Hereinafter the State Russian Museum (SRM).

in keeping with the taste of clients from St. Petersburg, whereas enamel icons that sold especially well in Rostov and other provincial towns were closer to folk art and the local pictorial tradition. To make the latter icons, enamellers followed the old icon painting instructions or copied contemporary popular religious pictures. Using a stencil to transfer the original onto a plaque, enamellers outlined the design and then coloured it with paints[12]. Rostov enamels were in great demand both in the capital and in remote Russian towns.

In the second half of the 19th century Rostov enamellers expanded their range of genres. When there appeared engravings with panoramas of cities and sights of local churches and monasteries, Rostov enamellers picked up the theme. At first single copies were made to serve as commemorative gifts, but subsequently enamel plaques with a panorama of Rostov became commonplace.

Enamel portraiture also emerged as an independent genre. Rostov enamellers early showed interest in the personality. Striking individualities and a multitude of psychological types and characteristics could be seen in the miniature images of saints made by Moshchansky, Rykunin and others. Arseny Vereshchagin, the Metropolitan of Yaroslavl, was known to have had his enamel portrait made possibly by a Rostov enameller[13] in the late 18th century. However, it was not until the second half of the 19th century that that genre was developed in earnest, when craftsmen less tied up with the church started to take commissions from the townsfolk. Despite their dependence on engravings from pictorial models or on daguerreotypes, Rostov enamellers charted new trends which were to be developed extensively in the future.

The Rostov enamelling business retained integrity and originality until the 1880s. The expanding market coupled with the competition mounted by the industrially produced icons caused the need to speed up the work process, which could not but affect the artistic quality of products. Division of labour and the serial production of the best examples of Rostov enamels sapped the style, turning an icon image into a token. Fine craftsmanship went downhill, and the industry gradually declined.

However, there remained craftsmen who passed the rich experience of the preceding generations to their younger colleagues in the late 19th century. Thanks to the efforts of the local inhabitants and the merited enameller Alexander Nazarov an enamelling class opened at the municipal vocational school, later transformed into a model enamel workshop. It was then that new ways were mapped out for the development of Rostov

[12] *Shchitova L. A.*, On the History of Rostov Enamels of the 18th—19th Centuries. Old Russian and Folk Art: Reports of the Zagorsk Museum Preserve, Moscow, 1990, p. 105.

[13] Arseny Vereshchagin's Diary (1786-1799), Vol. 2. Entry of July 15, 1797.

enamels with the help of Sergei Chekhonin, a professional artist who supervised the business. A fine connoisseur of applied arts, Chekhonin realized the importance of preserving the uniqueness of Rostov enamels, on the one hand, and of turning out modern products, on the other. On his initiative the genres of portraiture and cityscapes, which had sunk into oblivion by that time, were resumed. Enamellers also evolved samples of jewelry decorated with floral ornaments, which were in great demand. A new stage in the history of the business set in with the appearance of an artel, which transformed into the Rostov Finift Factory in 1959.

After the 1917 revolution Rostov's craftsmen formed an artel, which produced enamel caskets, boxes, brooches and cuff-links. Enamellers covered the shining surface of enamel plaques with floral designs, portraits of public and political leaders and agitprop symbols.

In their floral designs, Rostov enamellers employed techniques akin to those used in painting on porcelain. Enamel was of low grade at that time, and the paints employed were rather dull. The leading enamellers, among them Alexander Nazarov, Nikolai Karasev, Anna Yevdokimova and Victor Gorsky, nevertheless maintained the best qualities of Rostov enamels, depicting with great flair delicate field flowers and exuberant bouquets against a white enamel.

It proved less easy to retain the decorativeness of Rostov enamels in portraiture and miniatures, which became dominated by agitprop art of the period. Nonetheless, Alexander Nazarov and Nikolai Dubkov accomplished a great deal in that field, winning numerous awards at nationwide and international exhibitions. But the theory of easel painting as interpreted at that time and the infatuation with the agitprop poster idiom had an adverse effect on the pictorial tradition of Rostov enamels.

A passion for brush techniques developed by Rostov enamellers in imitation of oil painting was just as injurious. In the late 1950s, Maria Tone and Zinaida Zenkova, staff members of the Crafts Industries Research Institute, produced the first samples of enamel miniatures imitating oil paintings. Though the two women enamellers largely promoted the development of Rostov enamels and the training of craftsmen, the trends advocated by them in some measure disregarded the local tradition. The expressive means of Rostov enamels were again reduced to the minimum. Designs were based on a contrast between a painting in bright colours and a stark white or colour background. Conventional pictures were outlined with broad dabs and finished

with hatching. Beautiful miniature images again started to turn into schematic symbols. That desire to restrict Rostov enamels by some standards of a uniform style or alien traditions killed the original spirit of the craft. The innovations, however, failed to strike root due to either the enamellers' loyalty to their tradition or the Moscow instructors' sensitivity to the local tradition. Craftsmen increasingly turned to the sources of their business, studying the idiom of icon, fresco and enamel miniature painting. Thanks to the enthusiasm of local painters the old tradition and enamelling techniques were restored just as it already happened in the 19th century. Young enameller Alexander Alexeyev spent much time and efforts to restore the old enamelling process and to develop new paints that were in no way inferior to the old ones, drawing on old studies in glass-making and enamelling and on the recollections of local veterans.

Many talented painters capable of carrying on the local tradition joined the business in the late 1960s. They analyzed new possibilities for the development of Rostov enamels. Joining forces with jewelers Valentina Soldatova and Lydia Matakova, enamellers representing different generations produced original pieces of diverse shapes and purposes, decorated with enamel plaques with floral designs, genre paintings, landscapes or miniature portraits. Openwork or cloisons were most frequently employed in the decoration of these pieces. In Soldatova's works the openwork frame conformed to the integral rhythm of pictorial compositions, turning them into accomplished works of art. Matakova often preferred intricate filigree designs, seeking to convey the painter's idea both in the frame and the piece itself. The factory's leading jeweler today, Matakova is working hard to expand the range of local wares. Her frames and finished pieces have original shapes and filigree designs, perfectly harmonized with the local style of enamelling and the miniatures produced by different enamellers.

The factory's leading craftsmen are all highly original. Nikolai Kulandin, who belongs to the older generation, has a perfect command of the secrets of his craft. He succeeds in conveying the heroic mood and the lofty ideals of his favorite historical theme in enamel miniatures. A distinctive pictorial vocabulary has been used in his *Rostov Bells* (1967, RFF)[14], *Hunting* (1979, RFF) and *The Ice Lake Battle* (1978, RYMPAA)[15]. The influence of old Russian painting is felt in the flowing rhythm of lines and colour spots, in the strictly planned composition and in the peculiar treatment of figures. Kulandin remains true to the local tradition in his

[14] Hereinafter Rostov Finift Factory (RFF).

[15] Hereinafter Rostov-Yaroslavsky Museum Preserve of Architecture and Arts (RYMPAA)

portraits. Their range is fairly wide and includes famous figures of culture, heroes of Russian history and beautiful women's portraits imbued with noble spirit. The enameller's creativity was in no way restricted by the fact that he reproduced well-known portraits and in a way depended on an original drawing or painting. Thus working on Alexander Suvorov's portrait (1980, RYMPAA), Kulandin departed from the graphic original by Nikolai Utkin, treating conventionally the central figure of the decorative composition and showing military operations in the background. With the directness and enthusiasm of a folk artist, he went beyond physical likeness in a bid to give a detailed account of the great Russian military leader's heroic life.

Alexander Tikhov and Alexander Khaunov, graduates of the Fedoskino Art School, occupy a special place among the Rostov enamellers.

Tikhov glorified Russian nature in his lyrical landscapes, carrying on the theme masterfully developed by Mikhail Kulybin in the 1960s. Tikhov is in love with the beautiful scenery of Central Russia — town outskirts hidden in the garden greenery, abandoned villages and remote woods. Well-versed in Rostov's environs, he has the knack for choosing landscape motifs which acquire a special decorative effect on an enamel surface. Striving after integral and expressive compositions, Tikhov renounced the variegated palette and copious details. The tinged monochromatic colour scheme of his miniatures conveys the slightest change in the state of nature.

Khaunov's preferences are more diverse. He was equally skilled at landscapes, genre compositions and portraits. A fine sense of style and decorativeness helped him create exquisite and harmonious enamel miniatures. Together with Rostov's veteran enameller Ivan Soldatov, Khaunov extensively experimented with floral designs in the 1970s. They produced miniatures against white, black and colour backgrounds in the style of Moscow and Usolie enamels. Following strange traditions, the enamellers confidently transformed them. Khaunov was attracted by clear-cut graphic designs and consummately stylized ornamental motifs of Usolie enamels, while Soldatov exploited the orchestral colours of Moscow enamels. The two masters subsequently reverted to the traditional vocabulary of Rostov enamels, retaining their inimitable pictorial styles. Colour contrasts are combined with a certain expressive austerity in Khaunov's works. Soldatov's miniatures, on the contrary, look more tender and transparent: his lightened palette is better suited to convey the soft shapes of flowers and leaves.

The painters and jewelers of the older generation, including Kulandin, Khaunov, Soldatova and Matakova, were by rights awarded, in 1983, the Repin State Prizes for their work.

A close-knit group of gifted craftsmen appeared at the factory in the 1970s and the 1980s. Graduates of different art schools, skilled in the craft, mastered the secrets of the pictorial tradition and the decorative art of enamelling. The best of them, endowed with bright creative individuality, worked to renovate the idiom of Rostov enamels and to modify jewelry designs.

Boris Mikhailenko and Anatoly Zaitsev in their own way contributed to the development of portraiture. Their portrait gallery of heroes in the Patriotic War of 1812 presented personalities sharing a common sense of duty and love of their homeland. Zaitsev's laconic historical portraits are reminiscent in style of the classicist or romanticist models. Mikhailenko's portraits are richer in narrative detail. His characters are portrayed in their natural surroundings, finely attuned to the portrayed personality and his or her state of mind.

Lydia Samonova is also attracted by the same trend in portraiture and uses her gift of interpretative portrait in jewelry making. She decorates boxes, pendants and rings with her masterful portraits of famous personalities. Her caskets, made in collaboration with the young jeweller Mikhail Firulin, are bedecked with the portraits of Peter the Great and the Russian poet Nikolai Nekrasov and continue the old tradition of making snuff-boxes and other objects decorated with the portraits of donators.

Works by Alexander Alexeyev and Vladimir Grudinin reveal their interest in Russian history and local life. Alexeyev, who has introduced some innovative enamelling techniques, is also eagerly experimenting with genre paintings. His decorative panels *Old Masters* (1969, RFF) and *Grandfather's House* (1983, RYMPAA) show the past and present life of Rostov residents in complex compositions and poetic images. His enamels allow Alexeyev to produce a three-dimensional effect and to impart special luminescence to the air.

A graduate of the Kholui Art School, Grudinin took the Rostov enamelling tradition in his stride and enriched it with the sundry forms and colour schemes of Kholui lacquers. His works are multifarious — dynamic painting in the *Kulikovo Battle* triptych (the 1980s, RFF) and in the *Single Combat* (1985, RFF) and *Prince Igor* (1985, SRM) boxes gives way to the softly flowing lines and colours in the decoration of the *Date* glass-bead-box (1983, RFF). His generous palette floods the *Lake Nero* panel (1986, RFF) and the *Yaroslavl Tiles* box (1982, RFF) with kaleidoscopic colours. His floral designs are, on the contrary, characterised by an exquisite drawing and a restrained palette.

The beauty of their native land constantly inspires Rostov enamellers to mirror it in their landscapes. Alexander Khaunov, Victor Kotkov and Alexander Alexeyev are carrying on traditional panoramic landscape painting, showing the grandeur of Rostov's old Russian architecture. They create a generalised epic image of their «small homeland.»

Landscapes by Boris and Tatiana Mikhailenko convey a different lyrical mood. The two artists came to Rostov after graduating from the Moscow Art School and lost their hearts to the monumental beauty of old Russian architecture. Each of them has created a distinctive poetical image of Rostov, the town which has become so dear to both of them. A special feeling of Rostov's surroundings is what their works have in common. Tatiana Mikhailenko's *Twilight* miniature (1982, RYMPAA) reproduces the outlines of the Spaso-Yakovlevsky Monastery. Its cold silhouette seems to be coming out of the depths of the lake and dissolve in the warm beams of the setting sun. This illusory picture is firmly locked in an austere frame, which was done by Valentina Soldatova and which looks like a monastery wall. The painter's exalted brush and the jeweler's deft hands have preserved for us that inimitable instant of earthly beauty.

Valery Kochkin also has a fine feeling for Rostov landscapes and is very good at capturing their spirit in enamels. Using a wide range of pictorial techniques, he now raises the horizon, opening up the boundless expanses to the viewer, now covers the foreground with an openwork twining of tree branches.

The tone of his miniatures is echoed in Irina Alexeyeva's landscapes. Skilled at water-colour painting, Alexeyeva uses some of its techniques with great success in her enamel miniatures. Her landscapes and modest bouquets of field flowers and grasses seem to come alive on the translucent surface of painted enamels done by her husband, Alexander Alexeyev.

Floral motifs are central to Rostov enamels and are used in jewelry and household objects. All flowers and grasses of Russia, be it in bouquets or entwined into ornaments, generously fill the surface of enamel plaques. Every enameller who uses floral motifs has his or her distinctive style.

Elena Kotova, Natalia Serova and Vladimir Grudinin carefully model large-size garden flowers, forming of them posh bouquets or intricate ornamental compositions.

Elena Kotova is fond of elongated oval plaques with a gleaming surface of white enamel. She clusters flowers into thick bouquets, leaving much of the white background free. Colours tend to thicken towards the centre of the composition, with the motley colours fastened by the shadowed background under the bouquet. Elegant chrysanthemums and gorgeous roses combined with different grasses and other flowers look especially tangible and imbued with the freshness of living nature in Kotova's designs.

Natalia Serova fills the entire surface of enamel plaques with floral compositions in which bright silhouettes feature against coloured backgrounds. Fanciful ornamental inclusions make her painting intricate and especially decorative. Like Samonova, Serova is constantly looking for novel motifs and produces pieces unusual both in shape and floral design in collaboration with jewelers.

Grudinin's floral compositions, with their sophisticated colour schemes and refined drawing, look more reserved and closer to Tatiana Mikhailenko's and Valery Kochkin's miniatures in their emotional expressiveness.

These artists are attracted by the homely beauty of field flowers, berries and grasses. They have an unsurpassed ability to embody both the poetic images of awakening spring and the mystery of a summer night in their floral designs.

The outlines of exquisite floral designs are repeated in the curves of openwork frames which beautifully complement the enamel compositions. The new generation jewelers, such as Mikhail Firulin, Alexander Toporov, Alexander Serov, Sergei Lebedev and Sergei Sharabudinov vary the shapes of their products in a bid to perfect the decorative techniques. At their best they are co-authors on a par with enamellers.

The Rostov enamel business has an immense creative potential. There are quite a few gifted enamellers among the young painters, who produce original compositions. While embracing the local pictorial tradition, every new generation of artists imparts a new world outlook and their own idea of beauty to Rostov enamels. However, like any truly folk craft, Rostov enamels retain unchanged the ideal of the beauty of Russian nature and man.

Vocabulary of special terms

BARMY, a wide collar or shoulder-piece decorated with gold, pearls and gems, and sometimes with religious scenes painted in enamels on gold plaques.

CHALICE, a cup for the consecrated wine of the Eucharist.

CHARKA, a small cup for strong drinks.

CHASING, the tooling or surface modelling of metal to raise patterns in relief with hammer or chisel.

DAGUERREOTYPE, an early photograph made on a light-sensitive silver-coated metallic plate and developed by iodine vapour.

DEESIS, (from the Greek for «prayer») an Orthodox iconographic composition usually consisting of a range of the representations of saints, preceding Christ in their intercessional prayer for the people. The Deesis includes a representation of the Virgin and St. John the Baptist, the Apostles Peter and Paul, the Archangels Michael and Gabriel, as well as other saints placed in a certain order around Christ.

DROBNITSY, small mounts of various shape adorning household objects, church utensils and big icons with border scenes on separate plaques.

ENAMEL, (from the Latin «smaltum», meaning molten) a synonym to finift, denoting all types of enamel in West European countries. The term «enamel» became widespread in Russia in the 19th century, applied primarily to painted enamel miniatures, and gradually ousted the older term of «finift».

FINIFT, (from the Greek for «mix» or «fuse», according to some sources, translated as a «shining stone»), a term denoting all types of enamel in Ancient Greece, Byzantium and Russia. The term has survived in the name of the Rostov Finift Factory.

ICONOGRAPHY, a code of rule in Christian art to be followed in the representation of images of the Holy Writ.

SKAN', filigree used in jewelry since ancient times. Filigree objects consist of openwork ornaments made of thin gold, silver or copper twine, flat-beaten wire or thread.

ZERN', gold and silver filigree balls soldered upon the twine.

Illustrations

1. A cup. Late 17th century

The Birth of Finift

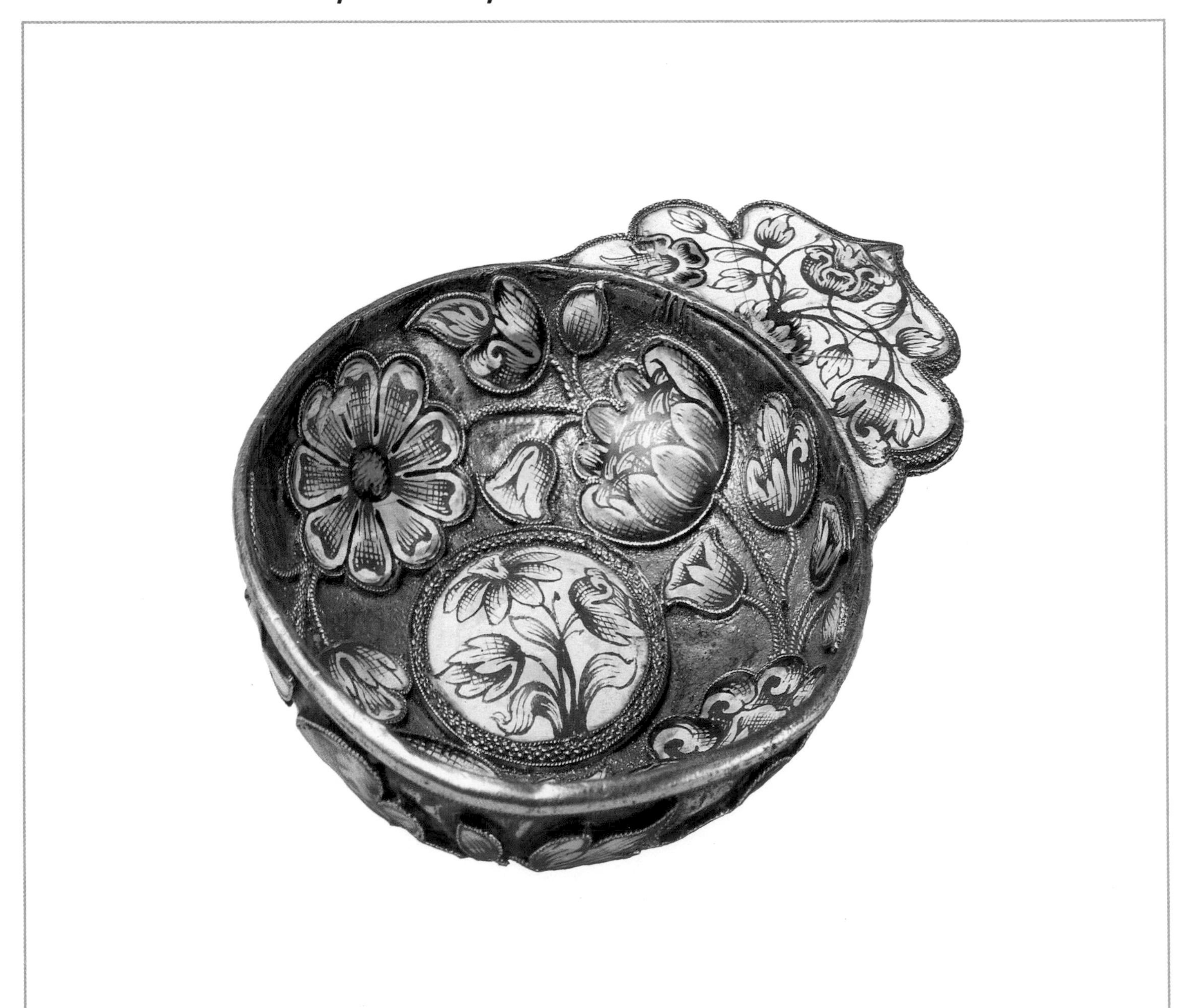

2. A Gospels corner mount
Late 17th century

3—4. Triptych. Deesis Range
The obverse and the reverse. 19th century
5. Triptych. Deesis Range with Chosen Saints. The reverse. 19th century

6. A Gospels cover. Third quarter of the 18th century

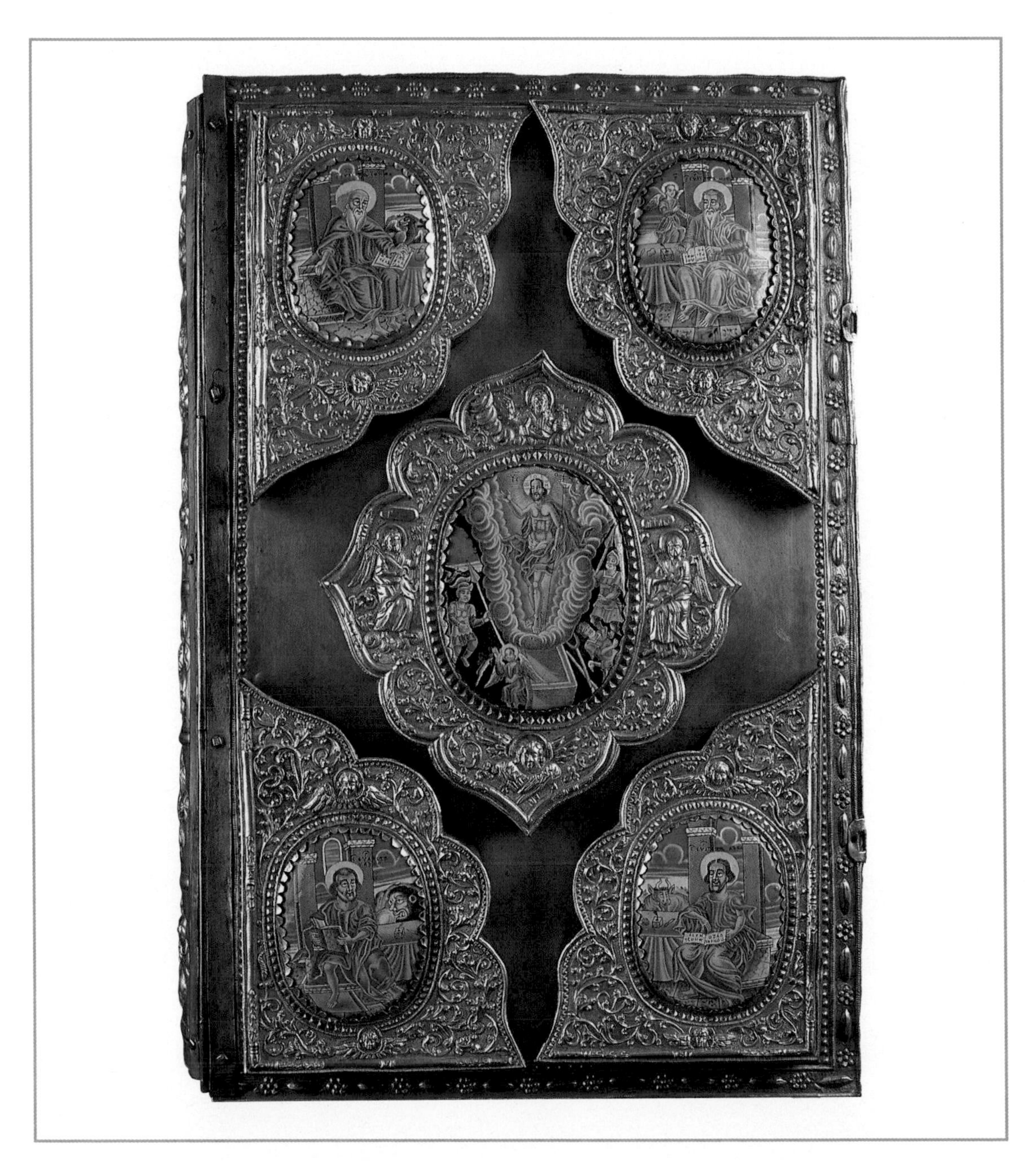

7. A heart-shaped icon *Crucifixion. The Descent of the Holy Ghost*
The obverse and the reverse
Mid-18th century

8. A plaque from a Gospels cover *Crucifixion*. Third quarter of the 18th century
9. A plaque from a Gospels cover *Resurrection*. Third quarter of the 18th century
10. *The Flagellation of Christ* icon. Third quarter of the 18th century

11. A plaque *Transfiguration*
Second half of the 18th century

12—15. Plaques showing the evangelists
Matthew, John, Luke and Mark. 18th century

16—17. Plaques from a mitre showing Christ and the Virgin
First half of the 19th century
18. A plaque showing St. John the Baptist
First half of the 19th century

19. A portrait of Catherine I. Second half of the 18th century
20. A portrait of Catherine II. Second half of the 18th century

21. A Badge of the Order of St. Anne. Mid-19th century
22. A Badge of the Order of St. Anne. Second half of the 19th century
23. A Badge of the Order of St. George of the fourth class. Late 19th century

24. A. I. Vsesvyatsky
A plaque from an altar cross
Crucifixion
with Entombment. 1796

25. A. I. Vsesvyatsky
A plaque from a processional cross
Resurrection and Descent into Hell. 1793

26. A. G. Moshchansky(?)
A plaque from an archimandrite's cross. 1777
27. An archimandrite's cross. 18th-19th centuries

28. A mitre. Late 1770s

29. A plaque from an archimandrite's cross. Late 1780s
30. An archimandrite's cross. 18th-19th centuries

31. *The Virgin Giving Joy to All Sorrowing* icon. Late 18th century
32. Master of the Rostov's Spaso-Yakovlevsky Monastery Milieu(?)
St. Dmitri of Rostov icon. Late 18th century
33. A. G. Moshchansky(?). *Rostov Wonder-Workers* icon. Early 19th century

34. An archimandrite's cross. 1791
35. Panagia *The Virgin of the Sign*. Circa 1793

36. A mitre. Late 18th-early 19th centuries
Fragment

37. Master of Rostov's Spaso-Yakovlevsky Monastery Milieu(?) A plaque *St. Dmitri of Rostov*. Late 18th-early 19th centuries
38. Master of Rostov's Spaso-Yakovlevsky Monastery Milieu(?) A plaque *St. John the Baptist*. Late 18th-early 19th centuries
39. Master of Rostov's Spaso-Yakovlevsky Monastery Milieu(?) A plaque *SS. Dmitri of Rostov And Nicholas the Wonder-Worker*. 1804

40. A mitre. Early 19th century
Fragment

41. A mitre. Late 18th-early 19th centuries
Two fragments

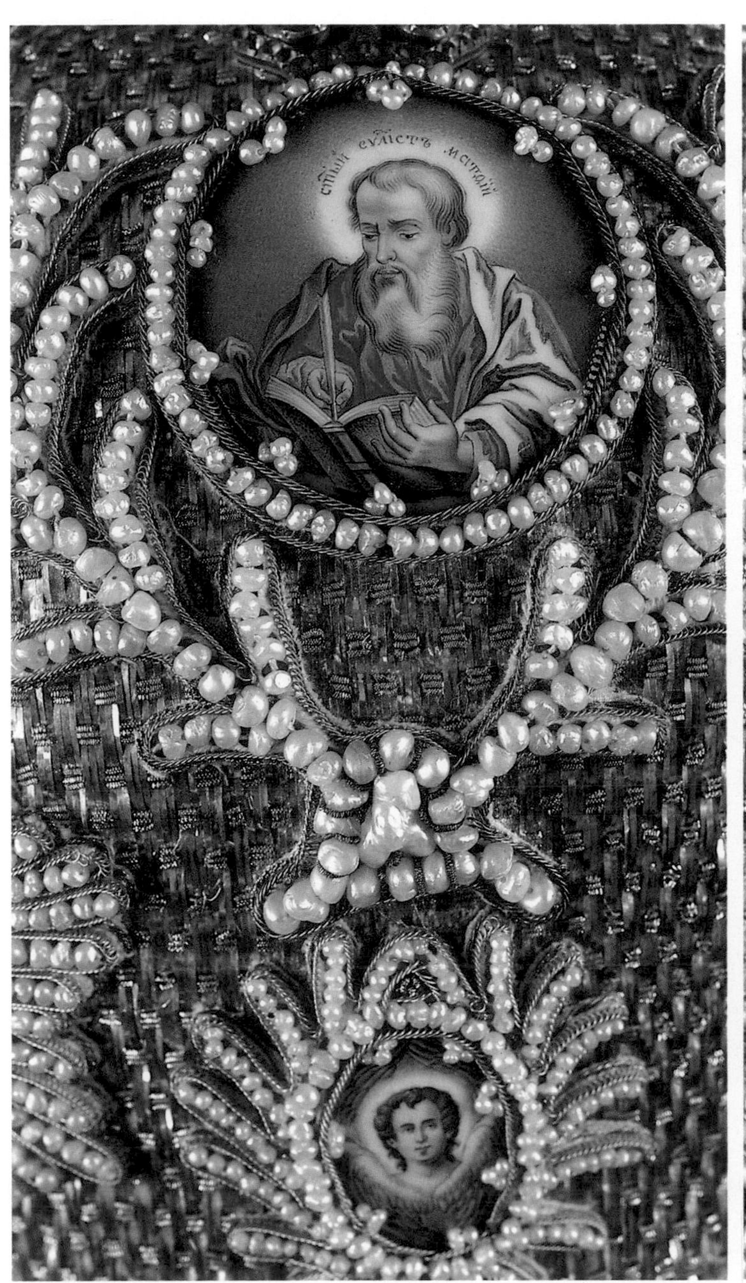

42. I. I. Shaposhnikov (?). A plaque *The New Testament Trinity*
First quarter of the 19th century

43. A. G. Tarasov. *SS. Constantin and Alexandra* icon
Mid-19th century

44. Y. I. Rykunin(?). A plaque from an icon *The Resurrection of Christ with Scenes from His Life Resurrection and Descent into Hell*. 1854

45. Y. I. Rykunin(?). A plaque from an icon *The Resurrection of Christ with Scenes from His Life*
The Last Supper. 1854

46. I. M. Zavyalov's workshop
Selected Saints icon. Mid-19th century

47. P. Eryomin
A plaque *St. Alexius Metropolitan of Moscow*. 1839
48. *The Virgin of the Sign* icon. First half of the 19th century
49. *St. Dmitri of Rostov* icon. First half of the 19th century

50. A plaque from a *Crucifixion* cross
First half of the 19th century

51. N. A. Salnikov(?)
The Synaxis of the Archangel Michael icon. 1876

52—53. Plaques from a Gospels cover showing the evangelists John and Mark
First half of the 19th century

54—55. Master *D. S.*
Plaques from the Holy Doors
The Virgin and *The Archangel Gabriel*. 1861

56. A Portrait of Iona Sysoyevich, Metropolitan of Rostov
Late 19th century

57. A Woman's Portrait. Second half of the 19th century
58. I. Shchennikov. A Portrait of an Unknown Man. 1849

59. *Rostov Wonder-Workers* hinged icon
Second half of the 19th century

60. *A Panorama of Rostov* plate
Early 20th century

61. A. A. Nazarov
The Virgin icon. 1912

62. Rostov's Model Enamel Workshop
A wooden treasure chest. 1910s

63. Rostov's Model Enamel Workshop
I. Pautov. A dish. 1916
64. Rostov's Model Enamel Workshop
K. Zherekhov. A box. 1914

64

65. A. A. Nazarov
Voroshilov at Manoeuvres
Early 1930s

66. N. A. Karasev
A woman's belt. 1957

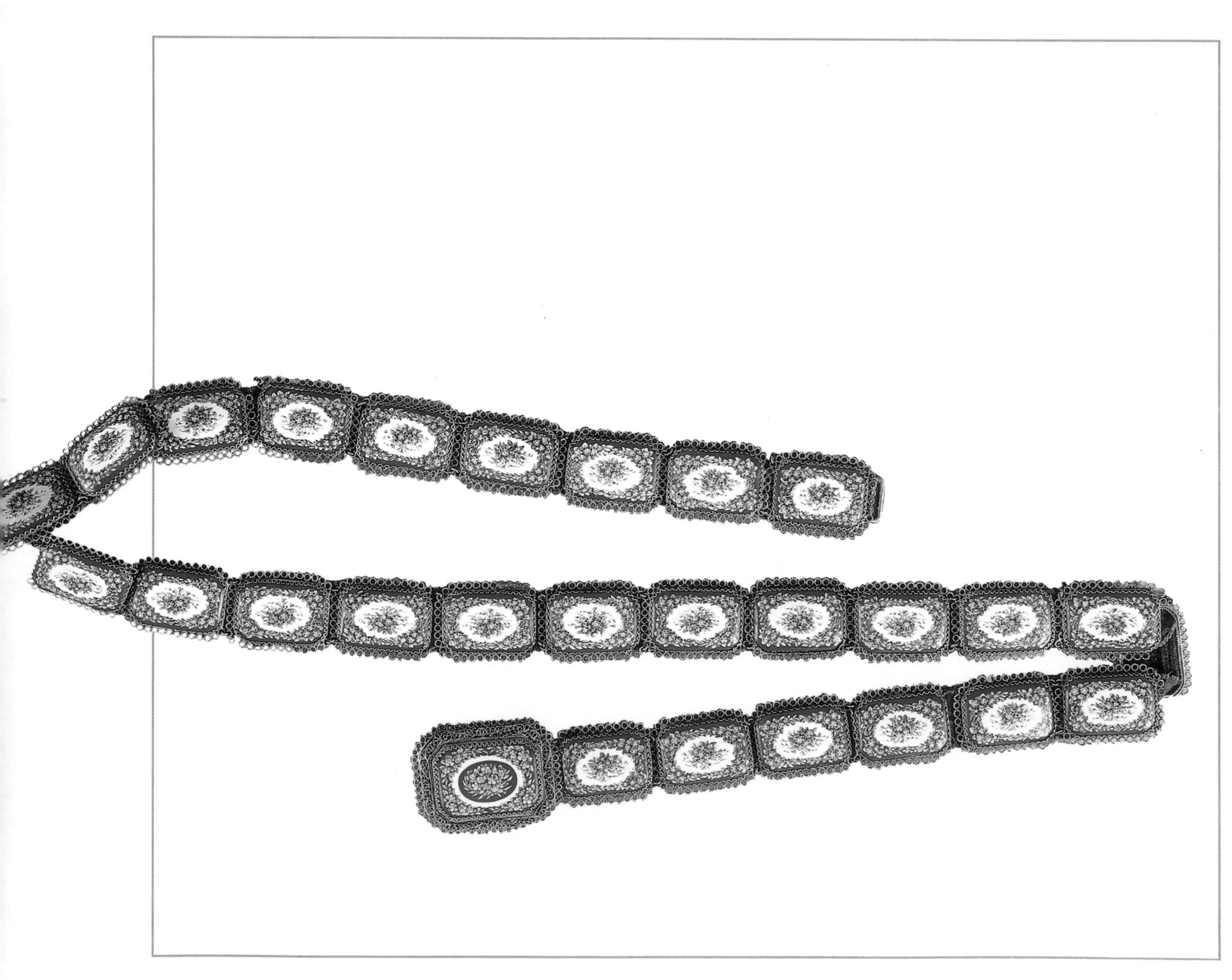

67. V. A. Odintsov
An octahedral box. 1956
68. M. M. Kulybin, M. A. Tone
A mirror with a landscape painting. Late 1950s

69—71. I. I. Soldatov, V. V. Soldatova
Flowers box. Late 1950s
Flowers casket. 1970s
Kalachi ear-rings. 1970s

72. A. M. Kokin
Beads. 1960s
73—74. I. I. Soldatov, V. V. Soldatova
Matryoshka Doll. Warrior. 1960s

75. V. D. Kotkov
Rostov the Great set. 1960s

76. N. A. Kulandin
Rostov Bells triptych. 1967

77—78. N. A. Kulandin, L. N. Matakova
Hunting box. 1979. *Pedlar* box. 1984
79. N. A. Kulandin, A. N. Bezugly
Mishenka, Dance! box. 1980

80. N. A. Kulandin, A. E. Zaitsev
Portrait of Alexander Suvorov. 1980
81. N. A. Kulandin
Portrait of Nikolai Nekrasov. 1982

82. A. A. Khaunov
Let's Defend the Russian Land triptych. 1980

83. A. A. Khaunov, L. N. Matakova
Decorative cup. 1978
84. A. A. Khaunov
Russian Motif box. 1977

85—86. I. I. Soldatov, V. V. Soldatova
Lily of the Valley brooch. 1973
Horse-shoe brooch. 1970s

87—88. I. I. Soldatov, V. V. Soldatova
Morning box. 1975. *Evening* set. 1978
89. E. S. Kotova, V. V. Soldatova
Brook bracelet. 1981

90. A. A. Khaunov, L. N. Matakova
Tile panel. 1978

91—92. A. G. Alexeyev
Glass-bead-box. 1979. *The Shining Sun* box. 1979
93. A. V. Tikhov, V. V. Soldatova
Rostov box. 1975

94—95. A. V. Tikhov, L. N. Matakova
Autumn panel. 1993
Towards Spring box. 1993

96—97. A. V. Tikhov, L. N. Matakova
Landscape panel. Late 1970s
Riverside Willow panel. 1980s

98. A. G. Alexeyev
Old Masters panel. 1969

99. A. G. Alexeyev, A. N. Bezugly. *Autumn in Osoyevo* box. 1980
100. A. G. Alexeyev, M. A. Firulin. *Boldino Autumn* box. 1980
101. B. M. Mikhailenko. *Alexander Pushkin* pendant

102. B. M. Mikhailenko, A. N. Bezugly. *Rostov the Great* box. 1981
103. B. M. Mikhailenko. *Yaroslavl Is 975 Years Old* box. 1985
104. T. S. Mikhailenko, V. V. Soldatova. *Twilight* box. 1980

105. B. M. Mikhailenko
Portrait of Nikolai Rayevsky. 1982
106. A. E. Zaitsev
Portrait of Dmitry Dokhturov, hero of the Patriotic War of 1812. 1985

107. V. D. Kochkin, M. A. Firulin. *The First Snow* panel. 1985
108. V. D. Kochkin, A. S. Serov. *Moon-lit Night* box. 1983
109. V. D. Kochkin, A. K. Toporov. *My Town* box. 1981

110. V. D. Kochkin, S. A. Lebedev
Rostov Laces box, 1985

111—112. E. M. Anisimova, V. A. Sharov
Rostov Sights box. 1980. *Autumn* box. 1980
113. V. I. Polyakov, L. N. Matakova
The Church of the Nativity of St. John the Baptist box. 1988

114. V. P. Grudinin, A. A. Vlasichev
Lake Nero panel. 1986

115—117. V. P. Grudinin, L. N. Matakova
Alyonushka box. 1980s
A Date glass-bead-box. 1983
Lel glass-bead-box. 1980s

118—119. V. P. Grudinin, L. N. Matakova
Fabulous set. 1988
Hawking box. 1986

120. V. P. Grudinin, A. N. Bezugly
The Kulikovo Battle triptych. 1981

121. L. D. Samonova, N. V. Serova, A. S. Serov. *The Seasons* flask. 1989
122—123. L. D. Samonova, M. A. Firulin
Peter the Great box. 1985
Medallion with a Portrait. 1989

124—125. E. S. Kotova, L. N. Matakova
Flora set. 1986
Pearly needle-case. 1984

126. N. V. Serova, L. N. Matakova
Flower Waltz box. 1990
127. N. V. Serova, V. E. Yakimov
Twilight set. 1993

128. T. S. Mikhailenko, V. V. Soldatova
Moscow Nights box. 1983
129. T. S. Mikhailenko, B. M. Mikhailenko
The Prize of the Graces set. 1984

130. V. D. Kochkin, M. A. Firulin
Evening box. 1987
131. V. P. Grudinin, A. K. Toporov
Gift box. 1985

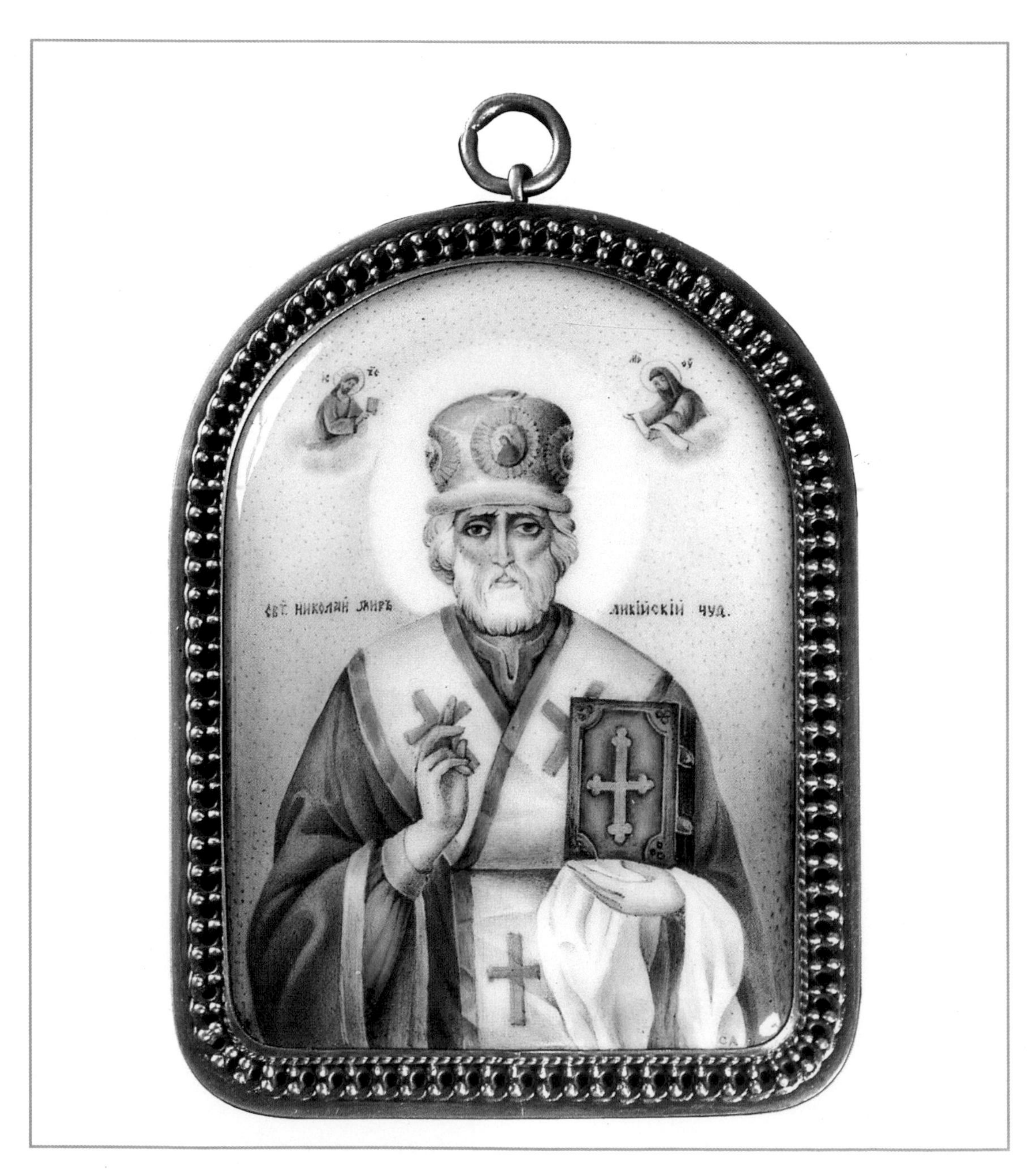

132. L. D. Samonova, M. A. Firulin
St. Nicholas the Wonder-Worker icon. 1994

133. L. D. Samonova, M. A. Firulin
St. Seraphin of Sarov icon. 1994

Portrait

134. Portrait
of General Bulatov,
hero of the Patriotic War
of 1812
Late 19th century

135. A. E. Zaitsev
Portrait of Mikhail Kutuzov. 1984

136—137. N. A. Kulandin
Portrait of Maria Volkonskaya. 1983
Portrait of Sergei Volkonsky. 1983

138. B. M. Mikhailenko
Portrait of Alexander Musin-Pushkin. 1986

139. B. M. Mikhailenko
Portrait of A. Titov. 1983

140. A. G. Alexeyev
Portrait of Mikhail Lomonosov. 1986

141. *A View of Rostov the Great* miniature
Second half of the 19th century

Landscapes

142. Rostov's Model Enamel Workshop
Rostov the Great plaque. 1910s
143. A. A. Nazarov. Ink-pot. 1915

144. V. A. Odintsov
Rostov the Great box. 1960s
145. A. A. Khaunov, L. N. Matakova
Rostov Domes plaque. 1981

146. N. A. Kulandin, M. A. Firulin. *Cathedral Square* box. 1990
147—148. N. A. Kulandin, L. N. Matakova
Rostov the Great box. 1991. *Yaroslavl Environs* plaque. 1983

149—151. A. A. Khaunov, L. N. Matakova
Morning. Rostov panel. 1991
Zagorsk on the Podol box. 1990
Moscow Kremlin box. 1992

152. V. D. Kochkin, A. K. Toporov. *My Town* box. 1981
153—154. A. A. Khaunov, L. N. Matakova
Pskov box. 1983. *Pavlovsk Park* box. 1985

155—156. A. A. Khaunov, L. N. Matakova
Morning Rostov panel. 1991
Native Town box. 1994

157—158. A. V. Tikhov, L. N. Matakova
Riverside Willow box. 1985. *Rural Landscape* panel. 1991

159. V. D. Kotkov, V. E. Yakimov
The Cathedral of Christ the Saviour box. 1989
160. V. D. Kotkov, A. K. Toporov
The Spaso-Yakovlevsky Monastery box. 1983

161—162. A. G. Alexeyev
Summer Night in Porechie panel. 1983
Grandfather's House panel. 1983

163. A. G. Alexeyev
Storm in Porechie panel. 1990s

164. A. G. Alexeyev
Spring Night box.

165—166. B. M. Mikhailenko
Karabikha box. 1981. *Yesenin* medallion. 1981

167—168. V. P. Grudinin, L. N. Matakova
The Church of Tolga in Rostov box. 1994
Rural Landscape box. 1994

169—171. V. D. Kochkin, A. K. Toporov
Autumn Rostov box. 1980
Autumn Mood box. 1980
Fairy-tale Town panel. 1978

172—175. V. D. Kochkin, V. S. Malenkin
Autumn Snow box. 1994
Landscape broochs. 1993. (Three)

176—177. I. A. Alexeyeva, A. G. Alexeyev
Spring box. 1993
Autumn box. 1993

178—179. I. A. Alexeyeva, I. A. Alexeyev
October brooch. 1990s
Autumn brooch. 1990s

180. V. V. Gorsky, V. V. Soldatova. *Bouquet* box. 1956
181. I. I. Soldatov, V. V. Soldatova. Brooch. 1950s
182. M. A. Tone, S. M. Karetnikova. Pendant. 1950s

Flowers of Russia

183. M. A. Tone. Brooch. 1950s
184. A. A. Khaunov, L. N. Matakova. *Evening* set. 1993

185. T. S. Mikhailenko, L. N. Matakova
Kokoshnik necklace. 1986

186. T. S. Mikhailenko, V. V. Soldatova
August box. 1979

187. E. S. Kotova, V. V. Soldatova. *Brook* bracelet. 1981
188. N. V. Serova, N. N. Mishin. *Expectation* flask. 1989
189. N. V. Serova, V. S. Malenkin. Brooch with a Still Life. 1989

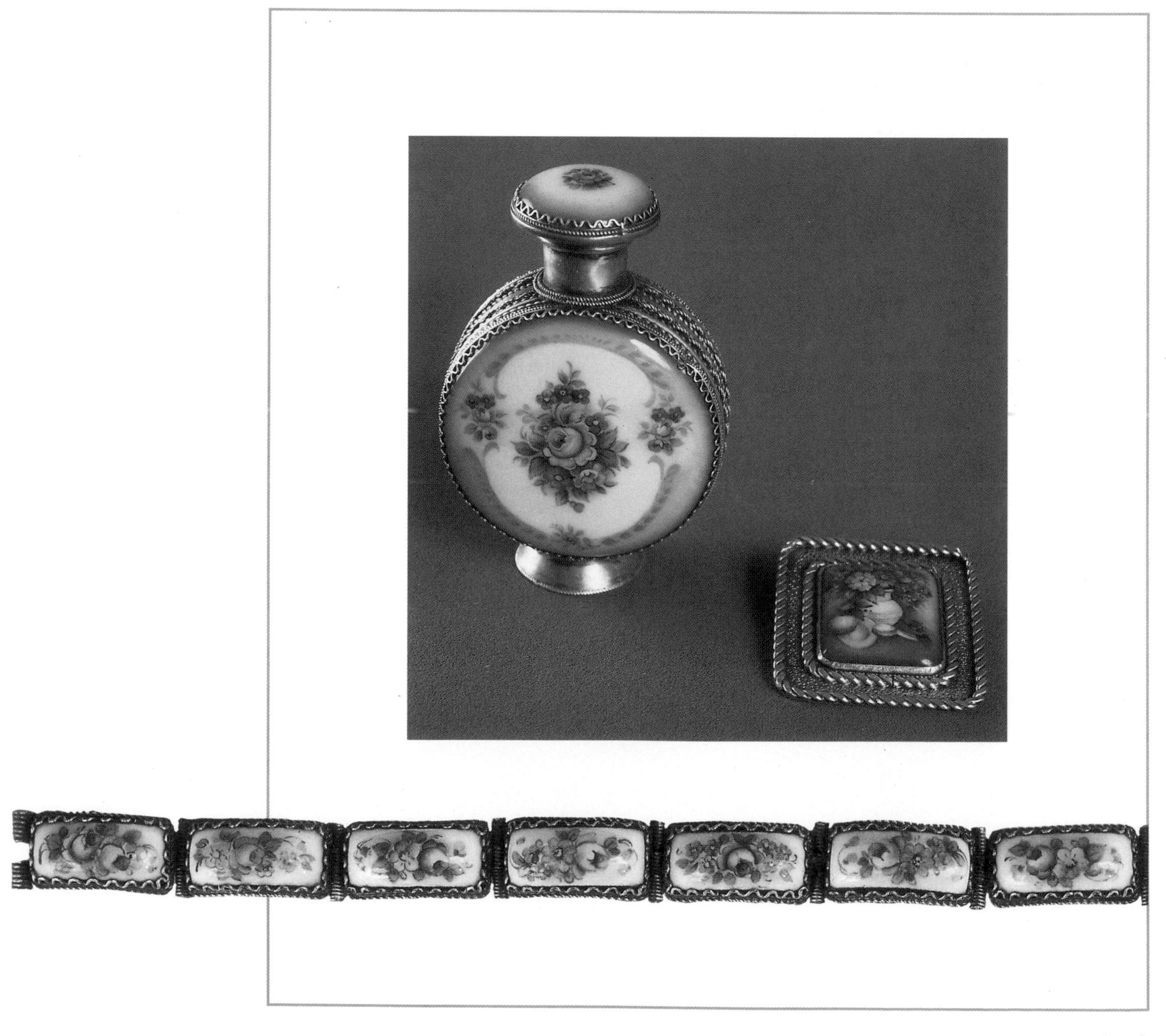

190. L. D. Samonova, M. A. Firulin. *Triumph* bracelet. 1987
191. L. D. Samonova, A. V. Mukharev. *Lilac* set. 1993

192—193. V. P. Grudinin, L. N. Matakova
Forest Glade box. 1989
Rostov Melodies set. 1989

194. N. A. Kulandin
Prince Vasilko panel. 1962

195. N. A. Kulandin
Battle. Alexander Nevsky panel. 1978

196. A. V. Tikhov, V. V. Soldatova
In the Meadow panel. 1977

197. B. M. Mikhailenko
Morozko panel. 1976

198. N. A. Kulandin, L. N. Matakova
September box. 1993

199. A. V. Tikhov, L. N. Matakova
Winter Motifs box. 1993

200. V. D. Kochkin, A. K. Toporov
Rostov Fairy-tale casket. 1982

201. A. A. Khaunov, A. K. Toporov
Rostov. Stables box. 1987

202. V. P. Grudinin, A. A. Vlasichev. *Bidding Farewell* folder. 1982
203. V. D. Kochkin, A. S. Serov. *Linden Bark Basket* box. 1987

204. L. D. Samonova, M. A. Firulin
Irina set. 1992

205. V. D. Kochkin, S. A. Lebedev. *Autumn Melody* box. 1989
206. V. D. Kochkin, A. S. Serov. *Soft Snow* brooch. 1989

207. N. V. Serova, A. S. Serov. *Morning Bouquet* box. 1984
208. S. Y. Pavlova, A. V. Mukharev. *Enchantment* box. 1993

209. L. D. Samonova, V. M. Kuznetsov
Rowan-tree set. 1993

210. L. D. Samonova, S. D. Sharabudinov
Miss Russia-1993 Crown. 1993

Catalogue

1. A cup. Late 17th century. Solvychegodsk
Silver, forging, twine, gilding; painted enamels
H. 2.4, diam. 7.2
Transferred from M. P. Botkin's collection in 1926.
SRM, inventory No. BK-3144

2. A Gospels corner mount. Late 17th century. Solvychegodsk. Inscribed in half-running hand:
(EVANGELIST LUKE; PAINTED BY AN ICON PAINTER OF USOLIE)
Silver, twine, gilding; painted enamels
20.6x14.6
Transferred from M. P. Botkin's collection in 1926.
SRM, inventory No. BK-3141

3. Triptych. Deesis Range. The obverse. 19th century
Copper, casting, enamelling
15x39x0.8
Transferred in 1968
SRM, inventory No. M-998

4. Triptych. Deesis Range. The reverse. 19th century
Copper, casting, enamelling
15x39x0.8
Transferred in 1968
SRM, inventory No. M-998

5. Triptych. Deesis Range with Chosen Saints. The reverse. 19th century
Copper, casting, enamelling
6.5x17x1.3
Transferred from an expedition to the Arkhangelsk Region
SRM, inventory No. M-873

6. A Gospels cover. Third quarter of the 18th century. Moscow
Copper, chasing, painted enamels
49x31.5x7
Transferred from Rostov's Spaso-Yakovlevsky Monastery.
RYMPAA, inventory No. YaK-413

7. A heart-shaped icon *Crucifixion. The Descent of the Holy Ghost*. The obverse and the reverse.
Mid-18th century. Moscow
Copper, painted enamels
6.5x5.5
SRM, inventory No. R-727

8. A plaque from a Gospels cover *Crucifixion*.
Third quarter of the 18th century. Moscow
Copper, painted enamels
6.6x5.8
Transferred from A. F. Kalikin's collection in 1956
SRM, inventory No. R-1868

9. A plaque from a Gospels cover *Resurrection*
Third quarter of the 18th century. Moscow
Copper, painted enamels
6.7x5.8
Transferred from A. F. Kalikin's collection in 1956
SRM, inventory No. R-1871

10. *The Flagellation of Christ* icon.
Third quarter of the 18th century. Moscow
Copper, painted enamels, metal frame, glass
8.7x7.7
SRM, inventory No. R-362

11. *Transfiguration* plaque.
Second half of the 18th century. Moscow
Copper, painted enamels, metal frame
11.5x9
SRM, inventory No. R-451

12—15. Plaques showing the Evangelists Matthew, John, Luke and Mark. 18th century. Kiev (?)
Copper, painted enamels, copper frame and plate
12x9.5
SRM, inventory No. R-2585, 2586, 2587, 2588

16—17. Plaques from a mitre showing Christ and the Virgin
First half of the 19th century. St. Petersburg
Copper, painted enamels, metal frame
6.6x5.4; 6.5x5.5
SRM, inventory No. R-325, 555

18. A plaque showing St. John the Baptist
First half of the 19th century. St. Petersburg
Copper, painted enamels
7.1x5.9
Transferred from A. F. Kalikin's collection in 1956
SRM, inventory No. R-1872

19. A Portrait of Catherine I
Second half of the 18th century. St. Petersburg
Copper, enamel miniature
8.2x6.6 (oval)
Transferred from the Hermitage in 1930
SRM, inventory No. Zh-3025

20. A Portrait of Catherine II
Second half of the 18th century. St. Petersburg
Copper, enamel miniature
8.2x6.7 (oval)
Transferred from the Hermitage in 1930
SRM, inventory No. Zh-3027

21. A Badge of the Order of St. Anne
Mid-19th century. St. Petersburg
Gold, enamelling
7.3x4.4
Transferred in 1969
SRM, inventory No. 5883

22. A Badge of the Order of St. Anne
Second half of the 19th century. St. Petersburg
Gold, enamelling
4.9x4.4 (with an ear)
Transferred from S. S. Zhuleva through a purchasing commission in 1988
SRM, inventory No. 19526

23. A Badge of the Order of St. George of the fourth class. Late 19th century. St. Petersburg
Gold, enamelling
3.5x3.5
SRM, inventory No. 2476 (BK-6-zol.)

24. Alexei Ignatievich Vsesvyatsky (1762-1831), Rostov enameller
A plaque from an altar cross
Crucifixion with Entombment. 1796. Rostov
Copper, silver, gilding, painted enamels
14x11
Transferred from Rostov's Church of All Saints in 1930
RYMPAA, inventory No. F-2301

25. Alexei Ignatievich Vsesvyatsky (1762-1831)
A plaque from a processional cross. *Resurrection and Descent into Hell*. 1793. Rostov
Copper, painted enamels
16.5x12.5
Transferred from Rostov's Cathedral of Assumption
RYMPAA, inventory No. F-21

26. Alexander Grigorievich Moshchansky(?) (1745-1824), staff votary at the Spaso-Yakovlevsky Monastery of Rostov, icon painter, artist and enameller
A plaque from an archimandrite's cross. 1777. Rostov
Inscribed on the reverse: *AM* and *1777*
Copper, painted enamels
7.4x5.5
Transferred from Rostov's Spaso-Yakovlevsky Monastery in 1923
RYMPAA, inventory No. F-2304

27. An archimandrite's cross. 18th-19th centuries. Rostov
Silver, chasing, gems, glass, painted enamels
15.5x8.5
Transferred from Rostov's Spaso-Yakovlevsky Monastery.
RYMPAA, inventory No. F-477

28. A mitre. Late 1770s. Rostov
Glazed brocade, velvet, pearls, glass, enamels
22x22
Transferred from Rostov's Spaso-Yakovlevsky Monastery.
RYMPAA, inventory No. DM-33

29. A plaque from an archimandrite's cross. Late 1780s. Rostov
Copper, painted enamels
7.5x6.5
Transferred from Rostov's Spaso-Yakovlevsky Monastery in 1923
RYMPAA, inventory No. F-2291

30. An archimandrite's cross. 18th-19th centuries. Rostov
Silver, gilding, chasing, paste, copper, painted enamels
14.5x7.4
Transferred from Rostov's Spaso-Yakovlevsky Monastery
RYMPAA, inventory No. DM-246

31. *The Virgin Giving Joy to All Sorrowing* icon
Late 18th century. Rostov
Copper, painted enamels, metal frame
7.3x5.2
SRM, inventory No. R-569

32. Master of the Rostov's Spaso-Yakovlevsky Monastery(?) Milieu.
St. Dmitry of Rostov icon. Late 18th century
Copper, painted enamels, metal frame
6.8x5
SRM, inventory No. R-730

33. Alexander Grigorievich Moshchansky(?)
Rostov Wonder-Workers icon. Early 19th century. Rostov
Copper, painted enamels, metal frame
8.5x6.1
SRM, inventory No. R-732

34. An archimandrite's cross. 1791. Rostov
Inlayed inscription: Donated by sexton Petr Ivanov of the Saviour Church on December 6, 1791.
Silver, painted enamels
9.1x6.5
Transferred from Rostov's Church of the Saviour on the Square in 1925
RYMPAA, inventory No. F-2293

35. Panagia *The Virgin of the Sign*. Circa 1793. Rostov
Inlayed inscription: Donated by sexton Petr Ivanov of the Saviour Church on November 27, 1793.
Silver, copper, painted enamels
9x7
Transferred from Rostov's Church of the Saviour on the Market Square in 1925
RYMPAA, inventory No. F-2294

36. A mitre. Late 18th-early 19th centuries. Fragment
Velvet, fringe, pearls, sequins, glass, painted enamels
H. 23, diam. 20
Transferred from Rostov's Spaso-Yakovlevsky Monastery
RYMPAA, inventory No. DM-35

37. Master of Rostov's Spaso-Yakovlevsky Monastery (?) Milieu
St. Dmitry of Rostov plaque. Late 18th-early 19th centuries
Copper, painted enamels
6.4x4.7
SRM, inventory No. R-731

38. Master of Rostov's Spaso-Yakovlevsky Monastery (?) Milieu.
St. John the Baptist plaque. Late 18th-early 19th centuries
Copper, painted enamels
6.5x4.7
SRM, inventory No. R-35

39. Master of Rostov's Spaso-Yakovlevsky Monastery (?) Milieu
SS. Dmitry of Rostov And St. Nicholas the Wonder-Worker plaque. 1804
Copper, painted enamels
7.2x5
SRM, inventory No. R-768

40. A mitre. Early 19th century. Fragment
Glazed brocade, wire-ribbon, sequins, flat-beaten wire, pearls, glass, amethysts, painted enamels
H. 21, diam. 19
Transferred from Rostov's Spaso-Yakovlevsky Monastery
RYMPAA, inventory No. T-2994

41. A mitre. Late 18th-early 19th centuries
Two fragments
Velvet, pearls, sequins, copper, painted enamels
23x18
Transferred from the Avraamiev Monastery in 1921
RYMPAA, inventory No. DM-34

42. Ivan Ivanovich Shaposhnikov (?), enameller of the first half of the 19th century and owner of an enamel business in Rostov
The New Testament Trinity plaque. First quarter of the 19th century
Copper, painted enamels, metal frame
7.5x14.6
Transferred from M. N. Tyunina in 1987
RYMPAA, inventory No. F-1977

43. Alexei Gavrilovich Tarasov, foreman enameller of the second third of the 19th century
SS. Constantin and Alexandra icon.
Mid-19th century. Rostov
Copper, painted enamels, metal frame
15.4x12.4
SRM, inventory No. R-834

44. Yakov Ivanovich Rykunin(?) (1816—1855), foreman enameller
A plaque from an icon *The Resurrection of Christ with Scenes from His Life*. 1854. Rostov
Resurrection and Descent into Hell
Copper, painted enamels, metal frame
23.8x20
Transferred from L. K. Markova through a purchasing commission in 1958
SRM, inventory No. R-2101

45. Yakov Ivanovich Rykunin(?) (1816—1855)
A plaque from an icon *The Resurrection of Christ with Scenes from His Life*. 1854. Rostov
The Last Supper
Copper, painted enamels
14x9.5
Transferred from L. K. Markova through a purchasing commission in 1958
SRM, inventory No. R-2081

46. Workshop of Ivan Matveyevich Zavyalov, framer and owner of an enamel workshop in Rostov in the first half of the 19th century
Selected Saints icon. Mid-19th century
Copper, painted enamels
21.2x17.9
SRM, inventory No. R-771

47. Petr Eryomin, enameller of the first half of the 19th century
A plaque *St. Alexius Metropolitan of Moscow*. 1839. Rostov
Inscription on the reverse: MASTER RO. UYE. NA. VO. BO. SLOBODA PEASANT PETR ERYOMIN July 14, 1839
Copper, painted enamels
Diam. 8.6
Transferred from S. I. Sheptunova through a purchasing commission in 1958
SRM, inventory No. R-2039

48. *The Virgin of the Sign* icon. First half of the 19th century. Rostov
Copper, painted enamels, metal frame
7.9x5.5
SRM, inventory No. R-597

49. *Dmitry of Rostov* icon. First half of the 19th century. Rostov
Copper, painted enamels, metal frame
8x6.2
SRM, inventory No. R-754

50. A plaque from a *Crucifixion* cross
First half of the 19th century. Rostov
Silver, glass, painted enamels
7.5x7.1
Transferred from Rostov's Church of the Saviour on the Market Square in 1925
RYMPAA, inventory No. F-2372

51. Nikolai Andreyevich Salnikov(?), enameller of the mid and second half of the 19th century.
Graduated from the St. Petersburg Academy of Arts in 1855 with the title of an artist without rank
The Synaxis of the Archangel Michael icon. 1876
Copper, painted enamels, metal frame
23x18.5
Transferred from Vinogradova in 1922
RYMPAA, inventory No. Ts-922/737

52—53. Plaques from a Gospels cover showing the Evangelists John and Mark. First half of the 19th century
Copper, painted enamels, metal frame
6.5x5.5
Transferred from Rostov's Church of the Exaltation of the Cross
RYMPAA, inventory No. F-456, F-455

54—55. Master *D. S*
Plaques from the Holy Doors. 1861. Rostov
The Virgin and *The Archangel Gabriel*
Copper, painted enamels, metal frame
18.3x14.6
Transferred from the Lazarevskaya Church of Rostov
RYMPAA, inventory No. F-1424, F-2155

56. A Portrait of Iona Sysoyevich, Metropolitan of Rostov. Late 19th century. Rostov
Copper, painted enamels, metal frame
21x18
RYMPAA, inventory No. F-950

57. A Woman's Portrait. Second half of the 19th century
Copper, painted enamels
6.7x5.2x0.2
Transferred from the Bykovs-Morokuyevs family through a purchasing commission in 1988
SRM, inventory No. R-4642

58. Ivan Shchennikov
A Portrait of an Unknown Man (copied from a daguerreotype). 1849
Copper, painted enamels
8x7
RYMPAA, inventory No. F-2059

59. *Rostov Wonder-Workers* hinged icon.
Second half of the 19th century
Copper, painted enamels, metal frame
14x19
Transferred from A. A. Titov's collection
RYMPAA, inventory No. F-2278

60. *A Panorama of Rostov* plate. Early 20th century
Copper, painted enamels, metal frame
15x19
Transferred from the Handicrafts Museum in 1938
SRM, inventory No. R-1092

61. Alexander Alexeyevich Nazarov (1872-1947), enameller and one of the founders of the Model Enamel Workshop and later of the Rostov Enamels Artel
The Virgin icon. 1912
Copper, painted enamels, wooden frame
17.2x12.9
RYMPAA, inventory No. F-1432

62. Rostov's Model Enamel Workshop
A wooden treasure chest. 1910s
Wood carving, copper, painted enamels
17.5x40x20
Transferred from the Handicrafts Museum in 1938
SRM, inventory No. R-1108

63. Rostov's Model Enamel Workshop.
I. Pautov. A dish. 1916
Inscriptions: «Paint. Pautov I. 1916» and «M. E. W.»
Copper, painted enamels
Diam. 17.5
Transferred from the Handicrafts Museum in 1938
SRM, inventory No. R-1149

64. Rostov's Model Enamel Workshop
K. Zherekhov. A box. 1914
Copper, painted enamels
H. 8, diam. 10
Transferred from the Handicrafts Museum in 1938
SRM, inventory No. R-1143 a, b

65. Alexander Alexeyevich Nazarov (1872-1947)
Voroshilov at Manoeuvres. Early 1930s. Rostov
Copper, painted enamels, metal frame
20x17
RYMPAA, inventory No. F-2363

66. Nikolai Alexandrovich Karasev (1909-1971), enameller, A. A. Nazarov's disciple, worked at the Rostov Artel from 1955
A woman's belt. 1957
Copper, painted enamels, twine, gilding
L. 97.5
Transferred from the Rostov Finift Factory in 1966.
RYMPAA, inventory No. F-2099

67. Vladimir Alexandrovich Odintsov (1927-1993), enameller
An octahedral box. 1956
Copper, painted enamels, twine, gilding
8x8x3.5
RFF, inventory No. 126

68. Mikhail Mikhailovich Kulybin (1910-1986), enameller
Maria Alexandrovna Tone (b. 1922), Moscow jeweler, member of the Russian Artists' Union
A mirror with a landscape painting. Late 1950s
Copper, painted enamels, twine, gilding, mirror
RFF, inventory No. 333

69. Ivan Ivanovich Soldatov (b. 1917), enameller, winner of the Order of Lenin and the Order of the Red Banner of Labour
Valentina Vasilievna Soldatova (b. 1927), jeweler, winner of the Repin State Prize of the Russian Federation and of the Order of the Red Banner of Labour
Flowers box. Late 1950s
Copper, painted enamels, twine, gilding
10.5x5.5x3
RFF, inventory No. 197

70. Ivan Ivanovich Soldatov (b. 1917)
Valentina Vasilievna Soldatova (b. 1927)
Flowers casket. 1970s
Copper, painted enamels, gilding
6.5x3.5x1.5
RFF, inventory No. 923

71. Ivan Ivanovich Soldatov (b. 1917)
Valentina Vasilievna Soldatova (b. 1927)
Kalachi ear-rings, 1970s
Copper, painted enamels, twine, gilding
RFF, inventory No. 639

72. Anatoly Mikhailovich Kokin (1924-1960?), enameller, worked at the Rostov Enamel Artel from 1951
Beads. 1960s
Copper, painted enamels, gilding
RFF, inventory No. 523

73. Ivan Ivanovich Soldatov (b. 1917)
Valentina Vasilievna Soldatova (b. 1927)
Matryoshka Doll. 1960s
Copper, painted enamels, twine, filigree balls, silver plating, oxidizing
6.5x5x4.5
RFF, inventory No. 270

74. Ivan Ivanovich Soldatov (b. 1917)
Valentina Vasilievna Soldatova (b. 1927)
Warrior. 1960s
Copper, painted enamels, twine, filigree balls, silver plating, oxidizing
4.5x10.5
RFF, inventory No. 248

75. Viktor Dmitrievich Kotkov (b. 1933), enameller
Rostov the Great set (mirror, brooch, ear-rings and ring). 1960s
Copper, painted enamels
Transferred from the Rostov Finift Factory in 1982
RYMPAA, inventory No. F-1548

76. Nikolai Alexandrovich Kulandin (b. 1930), enameller, one of the leading masters of the Rostov Finift Factory, winner of the Repin State Prize of the Russian Federation and of the Order of the Red Banner of Labour, member of the Russian Artists' Union.
The Rostov Bells triptych. 1967
Copper, painted enamels, twine, filigree balls, silver plating, oxidizing
0.7x19x9
RFF, inventory No. 9

77. Nikolai Alexandrovich Kulandin (b. 1930)
Lydia Nikolayevna Matakova (b. 1937), jeweler, one of the leading masters of the Rostov Finift Factory, winner of the Repin State Prize of the Russian Federation, of the Order of the Red Banner of Labour and of the medal For Labour Prowess
Hunting box. 1979
Copper, painted enamels, twine, silver plating, oxidizing
5.5x4x1.5
RFF, inventory No. 924

78. Nikolai Alexandrovich Kulandin (b. 1930)
Lydia Nikolayevna Matakova (b. 1937)
Pedlar box. 1984
Copper, painted enamels, twine, silver plating, oxidizing
2.7x3.2x3.2
RFF, inventory No. 121

79. Nikolai Alexandrovich Kulandin (b. 1930)
Anatoly Nikolayevich Bezugly (b. 1952), jeweler, worked at the Rostov Finift Factory until 1982
Mishenka, Dance! box. 1980
Copper, painted enamels, twine, silver plating, oxidizing
3x4.5x4.5
RFF, inventory No. 154

80. Nikolai Alexandrovich Kulandin (b. 1930)
Anatoly Efimovich Zaitsev (b. 1949), jeweler and enameller, member of the Russian Artists' Union, chief artist of the Rostov Finift Factory from 1978 to 1982; worked at the RFF until 1982
A Portrait of Alexander Suvorov. 1980
Copper, painted enamels, twine, silver plating, oxidizing
11.5x8.5x1.5
RFF, inventory No. 92

81. Nikolai Alexandrovich Kulandin (b. 1930)
A Portrait of Nikolai Nekrasov. 1982
Copper, painted enamels, twine, silver plating, oxidizing
11x6x2
RFF, inventory No. 830

82. Alexander Alexeyevich Khaunov (b. 1945), enameller, winner of the Repin State Prize of the Russian Federation and of a Badge of Honour, member of the Russian Artists' Union, chief artist of the Rostov Finift Factory from 1977 to 1978
Let's Defend the Russian Land triptych. 1980
Copper, painted enamels, twine, toreutic work, silver plating, oxidizing
7.5x19x1.3
RFF, inventory No. 84

83. Alexander Alexeyevich Khaunov (b. 1945)
Lydia Nikolayevna Matakova (b. 1937)
Decorative cup. 1978
Copper, painted enamels, twine, filigree balls, silver plating, oxidizing
H. 7, diam. 10.3
Transferred from the author in 1978
RYMPAA, inventory No. F-3166

84. Alexander Alexeyevich Khaunov (b. 1945)
Lydia Nikolaevna Matakova (b. 1937)
Russian Motif box. 1977
Copper, painted enamels, twine, gilding
H. 1.5, diam. 5.5
Transferred from the factory in 1978
RYMPAA, inventory No. F-76

85. Ivan Ivanovich Soldatov (b. 1917)
Valentina Vasilievna Soldatova (b. 1927)
Lily of the Valley brooch. 1973
Copper, painted enamels, twine, silver plating, oxidizing
RFF, inventory No. 472

86. Ivan Ivanovich Soldatov (b. 1917)
Valentina Vasilievna Soldatova (b. 1927)
Horse-shoe brooch. 1970s
Copper, painted enamels, twine, silver plating, oxidizing
RFF, inventory No. 516

87. Ivan Ivanovich Soldatov (b. 1917)
Valentina Vasilievna Soldatova (b. 1927)
Morning box. 1975
Copper, painted enamels, twine, filigree balls, silver plating, oxidizing
3.2x5x5
RFF

88. Ivan Ivanovich Soldatov (b. 1917)
Valentina Vasilievna Soldatova (b. 1927)
Evening set. 1978
Copper, painted enamels, twine, silver plating, oxidizing
RFF

89. Elena Sergeyevna Kotova (b. 1948), enameller, member of the Russian Artists' Union
Valentina Vasilievna Soldatova (b. 1927)
Brook bracelet. 1981
Copper, painted enamels, twine, silver plating, oxidizing
RFF, inventory No. 794

90. Alexander Alexeyevich Khaunov (b. 1945)
Lydia Nikolayevna Matakova (b. 1937)
Tile panel. 1978
Copper, painted enamels, twine, silver plating, oxidizing
9.5x9.5x1.5
RFF, inventory No. 21

91. Alexander Gennadievich Alexeyev (b. 1940), painter, jeweler, experimentalist technician, member of the Russian Artists' Union
Glass-bead-box. 1979
Copper, painted enamels, twine
14x3x3
RFF, inventory No. 906

92. Alexander Gennadievich Alexeyev (b. 1940)
The Shining Sun box. 1979
Copper, painted enamels, twine
4x3x3
RFF, inventory No. 200

93. Alexander Vasilievich Tikhov (b. 1948), enameller
Valentina Vasilievna Soldatova (b. 1927)
Rostov box. 1975
Copper, painted enamels, twine
RFF, inventory No. 142

94. Alexander Vasilievich Tikhov (b. 1948)
Lydia Nikolayevna Matakova (b. 1937)
Autumn panel. 1993
Copper, painted enamels, silver plating, oxidizing
9x9.8
RFF

95. Alexander Vasilievich Tikhov (b. 1948)
Lydia Nikolayevna Matakova (b. 1937)
Towards Spring box. 1993
Copper, painted enamels, silver plating, oxidizing
5.5x5x2.5
RFF

96. Alexander Vasilievich Tikhov (b. 1948)
Lydia Nikolayevna Matakova (b. 1937)
Landscape panel. Late 1970s
Copper, painted enamels, metal frame
Diam. 11
RFF

97. Alexander Vasilievich Tikhov (b. 1948)
Lydia Nikolayevna Matakova (b. 1937)
Riverside Willow panel. 1980s
Copper, painted enamels, twine, silver plating, oxidizing
Diam. 11
RFF, inventory No. 1152

98. Alexander Gennadievich Alexeyev (b. 1940)
Old Masters panel. 1969
Copper, painted enamels, metal frame
25x19
RFF, inventory No. 13

99. Alexander Gennadievich Alexeyev (b. 1940)
Anatoly Nikolayevich Bezugly (b. 1952)
Autumn in Osoyevo box. 1980
Copper, painted enamels, twine, silver plating, oxidizing
6x5x2.5
RFF, inventory No. 835

100. Alexander Gennadievich Alexeyev (b. 1940)
Mikhail Alexandrovich Firulin (b. 1956), jeweler
Boldino Autumn box. 1980
Copper, painted enamels, twine, silver plating, oxidizing
5.7x5.7x2.2
RFF, inventory No. 183

101. Boris Mikhailovich Mikhailenko (b. 1941), enameller and jeweler, member of the Russian Artists' Union, chief artist of the Rostov Finift Factory from 1970 to 1972, worked at the factory until 1987
Alexander Pushkin pendant
Copper, painted enamels, metal
Author's collection

102. Boris Mikhailovich Mikhailenko (b. 1941)
Anatoly Nikolayevich Bezugly (b. 1952)
Rostov the Great box. 1981
Copper, painted enamels, twine, silver plating, oxidizing
3.5x3.5x3.8
RFF, inventory No. 171

103. Boris Mikhailovich Mikhailenko (b. 1941)
Yaroslavl Is 975 Years Old box. 1985
Copper, painted enamels, twine, silver plating, oxidizing
7x5.5x3.5
RFF, inventory No. 855

104. Tatiana Sergeyevna Mikhailenko (b. 1944), enameller, member of the Russian Artists' Union
Valentina Vasilievna Soldatova (b. 1927)
Twilight box. 1980
Copper, painted enamels, twine, silver plating, oxidizing
4x4.8x2.2
RFF, inventory No. 165

105. Boris Mikhailovich Mikhailenko (b. 1941)
Portrait of Nikolai Rayevsky. 1982
Copper, painted enamels, German silver, tombac
11.3x9.6 (oval)
Transferred from the author in 1982.
RYMPAA, inventory No. F-1536

106. Anatoly Efimovich Zaitsev (b. 1949)
Portrait of Dmitry Dokhturov, hero of the Patriotic War of 1812. 1985
Copper, painted enamels, German silver, tombac
8.7x7.5 (oval)
Transferred from the author in 1986
RYMPAA, inventory No. RF-1835

107. Valery Dmitrievich Kochkin (b. 1957), enameller, member of the Russian Artists' Union, worked at the factory until 1993
Mikhail Alexandrovich Firulin (b. 1956)
The First Snow panel. 1985
Copper, painted enamels, twine
10.2x8x0.9
RFF, inventory No. 831

108. Valery Dmitrievich Kochkin (b. 1957)
Alexander Sergeyevich Serov (b. 1954), enameller
Moon-lit Night box. 1983
Copper, painted enamels, twine, silver plating, oxidizing
4.6x2.5
RFF, inventory No. 858

109. Valery Dmitrievich Kochkin (b. 1957)
Alexander Kirillovich Toporov (b. 1951), jeweler, worked at the Rostov Finift Factory until 1991
My Town box. 1981
Copper, painted enamels, twine, silver plating, oxidizing
4x3x2.5
RFF, inventory No. 823

110. Valery Dmitrievich Kochkin (b. 1957)
Sergei Anatolievich Lebedev (1947-1993), jeweler
Rostov Laces box, 1985
Copper, painted enamels, German silver, twine
8x8x4.5
RYMPAA, inventory No. 1795

111. Elena Mikhailovna Anisimova (b. 1952), enameller
Victor Alexandrovich Sharov (b. 1955), jeweler
Rostov Sights box. 1980
Copper, painted enamels, twine, silver plating, oxidizing
5x4.5x4
RFF, inventory No. 157

112. Elena Mikhailovna Anisimova (b. 1952)
Victor Alexandrovich Sharov (b. 1955)
Autumn box. 1980
Copper, painted enamels, twine, silver plating, oxidizing
5x5x4
RFF, inventory No. 148

113. Victor Ivanovich Polyakov (b. 1955), enameller, chief artist of the Rostov Finift Factory from 1983 to 1991, worked at the factory until 1991
Lydia Nikolayevna Matakova (b. 1937)
The Church of the Nativity of St. John the Baptist box. 1988
Copper, painted enamels, twine, silver plating, oxidizing
3.8x4.8x2.3
RFF, inventory No. 1203

114. Vladimir Pavlovich Grudinin (b. 1954), enameller, member of the Russian Artists' Union
Alexander Alexandrovich Vlasichev (b. 1956), jeweler, worked at the factory until 1990
Lake Nero panel. 1986
Copper, painted enamels, metal frame
8.4x7.4x1.5
RFF, inventory No. 70

115. Vladimir Pavlovich Grudinin (b. 1954)
Lydia Nikolayevna Matakova (b. 1937)
Alyonushka box. 1980s.
Copper, painted enamels, twine, silver plating, oxidizing
4.1x2.3x1.9
RFF, inventory No. 971

116. Vladimir Pavlovich Grudinin (b. 1954)
Lydia Nikolayevna Matakova (b. 1937)
A Date glass-bead-box. 1983
Copper, painted enamels, twine, silver plating, oxidizing
4.1x2.3x1.9
RFF

117. Vladimir Pavlovich Grudinin (b. 1954)
Lydia Nikolayevna Matakova (b. 1937)
Lel glass-bead-box. 1980s.
Copper, painted enamels, twine, silver plating, oxidizing
4.1x2.3x1.9
RFF, inventory No. 565

118. Vladimir Pavlovich Grudinin (b. 1954)
Lydia Nikolayevna Matakova (b. 1937)
Fabulous set. 1988
Copper, painted enamels, twine, silver plating, oxidizing
RFF

119. Vladimir Pavlovich Grudinin (b. 1954)
Lydia Nikolayevna Matakova (b. 1937)
Hawking box. 1986
Copper, painted enamels, twine, silver plating, oxidizing
4x4x4.3
RFF, inventory No. 1092

120. Vladimir Pavlovich Grudinin (b. 1954)
Anatoly Nikolayevich Bezugly (b. 1952)
The Kulikovo Battle triptych. 1981
Copper, painted enamels, twine, silver plating, oxidizing
22x9.5
RFF, inventory No. 86

121. Larissa Dolgatovna Samonova (b. 1951), enameller, member of the Russian Artists' Union
Natalia Vladimirovna Serova (b. 1954), enameller, member of the Russian Artists' Union
Alexander Sergeyevich Serov (b. 1954), jeweler
The Seasons flask. 1989
Copper, painted enamels, twine, silver plating, oxidizing
RFF

122. Larissa Dolgatovna Samonova (b. 1951)
Mikhail Alexandrovich Firulin (b. 1956)
Peter the Great box. 1985
Copper, painted enamels, twine, silver plating, oxidizing
6.5x5.5x2.5
RFF, inventory No. 220

123. Larissa Dolgatovna Samonova (b. 1951)
Mikhail Alexandrovich Firulin (b. 1956)
Medallion with a Portrait. 1989
Copper, painted enamels, twine, silver plating, oxidizing
RFF

124. Elena Sergeyevna Kotova (b. 1948), enameller, member of the Russian Artists' Union
Lydia Nikolayevna Matakova (b. 1937)
Flora set (brooch and ear-rings). 1986
Copper, painted enamels, twine, silver plating, oxidizing
RFF

125. Elena Sergeyevna Kotova (b. 1948)
Lydia Nikolayevna Matakova (b. 1937)
Pearly needle-case. 1984
Copper, painted enamels, twine, silver plating, oxidizing
RFF

126. Natalia Vladimirovna Serova (b. 1954)
Lydia Nikolayevna Matakova (b. 1937)
Flower Waltz box. 1990
Copper, painted enamels, twine, filigree balls, silver plating, oxidizing
H. 3.5, diam. 7.5
RFF

127. Natalia Vladimirovna Serova (b. 1954)
Vyacheslav Evgenievich Yakimov (b. 1961), jeweler
Twilight set. 1993
Copper, painted enamels, twine, filigree balls, silver plating, oxidizing
RFF

128. Tatiana Sergeyevna Mikhailenko (b. 1944)
Valentina Vasilievna Soldatova (b. 1927)
Moscow Nights box. 1983
Copper, painted enamels, twine, silver plating, oxidizing
4.3x4.3x3.5
RYMPAA, inventory No. F-1656

129. Tatiana Sergeyevna Mikhailenko (b. 1944)
Boris Mikhailovich Mikhailenko (b. 1941)
The Prize of the Graces set (ear-rings, ring and pendant). 1984
Copper, painted enamels, metal frame
RYMPAA, inventory No. F-1692 (1-4)

130. Valery Dmitrievich Kochkin (b. 1957)
Mikhail Alexandrovich Firulin (b. 1956)
Evening box. 1987
Copper, painted enamels, twine, silver plating, oxidizing
6x5x2
RFF, inventory No. 1094

131. Vladimir Pavlovich Grudinin (b. 1954)
Alexander Kirillovich Toporov (b. 1951)
Gift box. 1985
Copper, painted enamels, twine, silver plating, oxidizing
6x8.5x2.7
RFF

132. Larissa Dolgatovna Samonova (b. 1951)
Mikhail Alexandrovich Firulin (b. 1956)
St. Nicholas the Wonder-Worker icon. 1994.
Copper, painted enamels, metal frame
RFF

133. Larissa Dolgatovna Samonova (b. 1951)
Mikhail Alexandrovich Firulin (b. 1956)
St. Seraphin of Sarov icon. 1994
Copper, painted enamels, metal frame
RFF

134. Portrait of General M. L. Bulatov, hero of the Patriotic War of 1812. Late 19th century
Copper, painted enamels, metal frame
10x8.5
Presented by I. I. Furtov in 1916
RYMPAA, inventory No. F-2060

135. Anatoly Efimovich Zaitsev (b. 1949)
Portrait of Mikhail Kutuzov. 1984
Copper, painted enamels, German silver
9.2x8.3
Transferred from the author in 1984
RYMPAA, inventory No. F-1715

136. Nikolai Alexandrovich Kulandin (b. 1930)
Portrait of Maria Volkonskaya. 1983
Copper, painted enamels, metal frame
8.7x7.4
Transferred from the author in 1983
RYMPAA, inventory No. F-1668

137. Nikolai Alexandrovich Kulandin (b. 1930)
Portrait of Sergei Volkonsky. 1983
Copper, painted enamels
9x7.7
Transferred from the author in 1983
RYMPAA, inventory No. F-1639

138. Boris Mikhailovich Mikhailenko (b. 1941)
Portrait of Alexander Musin-Pushkin. 1986
Copper, painted enamels, metal frame
10.7x9.5
RFF, inventory No. 795

139. Boris Mikhailovich Mikhailenko (b. 1941)
Portrait of A. Titov. 1983
Copper, painted enamels, German silver
10.3x8.8
Donated by the author in 1983
RYMPAA, inventory No. 1679

140. Alexander Gennadievich Alexeyev (b. 1940)
Portrait of Mikhail Lomonosov. 1986
Copper, painted enamels, German silver, brass, engraving
20.2x15.6
Transferred from the author in 1986
RYMPAA, inventory No. 1837

141. *A View of Rostov the Great* miniature
Second half of the 19th cent.
Copper, painted enamels
11.7x14
Received in 1918.
RYMPAA, inventory No. F-2374

142. Rostov's Model Enamel Workshop
Rostov the Great plaque. 1910s
Copper, painted enamels
6x8
SRM, inventory No. R-1091

143. Alexander Alexeyevich Nazarov (1872-1947)
Ink-pot. 1915
Copper, painted enamels
4.5x5.5x4
Transferred from the Handicrafts Museum in 1938
SRM, inventory No. R-1146 (a, b)

144. Vladimir Alexandrovich Odintsov (1927-1993)
Rostov the Great box. 1960s
Copper, painted enamels, twine
8x6x2
RFF, inventory No. 837

145. Alexander Alexeyevich Khaunov (b. 1945)
Lydia Nikolayevna Matakova (b. 1937)
Rostov Domes plaque. 1981
Copper, painted enamels, twine, silver plating, oxidizing
5.5x5.5x1.5
RFF, inventory No. 67

146. Nikolai Alexandrovich Kulandin (b. 1930)
Mikhail Alexandrovich Firulin (b. 1956)
Cathedral Square box. 1990
Copper, painted enamels, twine, silver plating, oxidizing
6x7x3
RFF, inventory No. 1096

147. Nikolai Alexandrovich Kulandin (b. 1930)
Lydia Nikolayevna Matakova (b. 1937)
Rostov the Great box. 1991
Copper, painted enamels, twine, silver plating, oxidizing
6.5x3.5x2
RFF

148. Nikolai Alexandrovich Kulandin (b. 1930)
Lydia Nikolayevna Matakova (b. 1937)
Yaroslavl Environs plaque. 1983
Copper, painted enamels, twine, silver plating, oxidizing
5.5x5
RFF, inventory No. 951

149. Alexander Alexeyevich Khaunov (b. 1945)
Lydia Nikolayevna Matakova (b. 1937)
Morning. Rostov panel. 1991
Copper, painted enamels, twine, silver plating, oxidizing
10x7.5
RFF

150. Alexander Alexeyevich Khaunov (b. 1945)
Lydia Nikolayevna Matakova (b. 1937)
Zagorsk on the Podol box. 1990
Copper, painted enamels, twine, silver plating, oxidizing
6x4.5x2.5
RFF, inventory No. 1098

151. Alexander Alexeyevich Khaunov (b. 1945)
Lydia Nikolayevna Matakova (b. 1937)
Moscow Kremlin box. 1992
Copper, painted enamels, twine, silver plating, oxidizing
5x7.5x1.8
RFF, inventory No. 190

152. Valery Dmitrievich Kochkin (b. 1957)
Alexander Kirillovich Toporov (b. 1951)
My Town box. 1981
Copper, painted enamels, twine, silver plating, oxidizing
4x3x2.5
RFF, inventory No. 823

153. Alexander Alexeyevich Khaunov (b. 1945)
Lydia Nikolayevna Matakova (b. 1937)
Pskov box. 1983
Copper, painted enamels, twine, silver plating, oxidizing
4.5x2.5x1.5
RFF, inventory No. 762

154. Alexander Alexeyevich Khaunov (b. 1945)
Lydia Nikolayevna Matakova (b. 1937)
Pavlovsk Park box. 1985
Copper, painted enamels, twine, silver plating, oxidizing
3x4x5
RFF, inventory No. 164

155. Alexander Alexeyevich Khaunov (b. 1945)
Lydia Nikolayevna Matakova (b. 1937)
Morning Rostov panel. 1991
Copper, painted enamels, twine, silver plating, oxidizing
10.5x7.5
RFF

156. Alexander Alexeyevich Khaunov (b. 1945)
Lydia Nikolayevna Matakova (b. 1937)
Native Town box. 1994
Copper, painted enamels, twine, silver plating, oxidizing
6x4.5x2.5
RFF

157. Alexander Vasilievich Tikhov (b. 1948)
Lydia Nikolayevna Matakova (b. 1937)
Riverside Willow box. 1985
Copper, painted enamels, twine, silver plating, oxidizing
3x4x7
RFF, inventory No. 1041

158. Alexander Vasilievich Tikhov (b. 1948)
Lydia Nikolayevna Matakova (b. 1937)
Rural Landscape panel. 1991
Copper, painted enamels, twine, silver plating, oxidizing
6.9x8.7
RFF

159. Victor Dmitrievich Kotkov (b. 1933), enameller
Vyacheslav Evgenievich Yakimov (b. 1961)
The Cathedral of Christ the Saviour box. 1989
Copper, painted enamels, twine, silver plating, oxidizing
H. 3.5, diam. 5.7
RFF, inventory No. 1127

160. Victor Dmitrievich Kotkov (b. 1933)
Alexander Kirillovich Toporov (b. 1951)
The Spaso-Yakovlevsky Monastery box. 1983
Copper, painted enamels, twine, silver plating, oxidizing
H. 2.6, diam. 4.2
RFF

161. Alexander Gennadievich Alexeyev (b. 1940)
Summer Night in Porechie panel. 1983
Copper, painted enamels, metal frame
Diam. 14.5
Received from the author in 1983
RYMPAA, inventory No. F-1657

162. Alexander Gennadievich Alexeyev (b. 1940)
Grandfather's House panel. 1983
Copper, painted enamels, metal frame
Diam. 18
Received from the author in 1983
RYMPAA, inventory No. F-1627

163. Alexander Gennadievich Alexeyev (b. 1940)
Storm in Porechie panel. 1990s
Copper, painted enamels, metal frame
From the author's collection

164. Alexander Gennadievich Alexeyev (b. 1940)
Spring Night box.
Painted enamels, German silver, twine
5.8x7.4x2.6
Received from the author in 1982
RYMPAA, inventory No. F-1529

165. Boris Mikhailovich Mikhailenko (b. 1941)
Karabikha box. 1981
Copper, painted enamels, German silver, metal frame
7.5x5.5x1.7
Received from the author in 1982
RYMPAA, inventory No. F-1520

166. Boris Mikhailovich Mikhailenko (b. 1941)
Yesenin medallion. 1981
Copper, painted enamels, German silver
4.2x2.5
Received from the author in 1982
RYMPAA, inventory No. F-1516

167. Vladimir Pavlovich Grudinin (b. 1954)
Lydia Nikolayevna Matakova (b. 1937)
The Church of Tolga in Rostov box. 1994
Copper, painted enamels, twine, silver plating, oxidizing
5.5x5.5x2.4
RFF

168. Vladimir Pavlovich Grudinin (b. 1954)
Lydia Nikolayevna Matakova (b. 1937)
Rural Landscape box. 1994
Copper, painted enamels, twine, filigree balls, silver plating, oxidizing
4.2x4.2x3
RFF

169. Valery Dmitrievich Kochkin (b. 1957)
Alexander Kirillovich Toporov (b. 1951)
Autumn Rostov box. 1980
Copper, painted enamels, twine, silver plating, oxidizing
6x4x2
RFF, inventory No. 177

170. Valery Dmitrievich Kochkin (b. 1957)
Alexander Kirillovich Toporov (b. 1951)
Autumn Mood box. 1980
Copper, painted enamels, twine, silver plating, oxidizing
6x3.5x2
RFF, inventory No. 176

171. Valery Dmitrievich Kochkin (b. 1957)
Alexander Kirillovich Toporov (b. 1951)
Fairy-tale Town panel. 1978
Copper, painted enamels, wood
10x8
RFF, inventory No. 978

172. Valery Dmitrievich Kochkin (b. 1957)
Valery Sergeyevich Malenkin (b. 1958), jeweler, worked at the factory until 1991
Autumn Snow box. 1994
Copper, painted enamels, twine, niello, silver plating, oxidizing
From the author's collection

173. Valery Dmitrievich Kochkin (b. 1957)
Valery Sergeyevich Malenkin (b. 1958)
Landscape brooch. 1993
Copper, painted enamels, metal frame
From the author's collection

174. Valery Dmitrievich Kochkin (b. 1957)
Valery Sergeyevich Malenkin (b. 1958)
Landscape brooch. 1993
Copper, painted enamels, metal frame
From the author's collection

175. Valery Dmitrievich Kochkin (b. 1957)
Valery Sergeyevich Malenkin (b. 1958)
Landscape brooch. 1993
Copper, painted enamels, metal frame
From the author's collection

176. Irina Anatolievna Alexeyeva (b. 1953), watercolour artist, enameller
Alexander Gennadievich Alexeyev (b. 1940)
Spring box. 1993
Copper, painted enamels, twine
From the author's collection

177. Irina Anatolievna Alexeyeva (b. 1953)
Alexander Gennadievich Alexeyev (b. 1940)
Autumn box. 1993
Copper, painted enamels, twine
From the author's collection

178. Irina Anatolievna Alexeyeva (b. 1953)
Ivan Alexandrovich Alexeyev (b. 1977), student
October brooch. 1990s
From the author's collection

179. Irina Anatolievna Alexeyeva (b. 1953)
Ivan Alexandrovich Alexeyev (b. 1977), student
Autumn brooch. 1990s
From the author's collection

180. Victor Vladimirovich Gorsky (1920-1971), enameller, worked at the Rostov Finift Factory from 1930 through the 1960s
Valentina Vasilievna Soldatova (b. 1927)
Bouquet box. 1956
Copper, painted enamels, twine, gilding
6.5x4.5x1.5
RFF, inventory No. 128

181. Ivan Ivanovich Soldatov (b. 1917)
Valentina Vasilievna Soldatova (b. 1927)
Brooch. 1950s
Copper, painted enamels, twine, gilding
RFF, inventory No. 502

182. Maria Alexandrovna Tone (b. 1922)
Sofia Mikhailovna Karetnikova (b. 1920)
Pendant. 1950s
Copper, painted enamels, twine, gilding
RFF, inventory No. 533

183. Maria Alexandrovna Tone (b. 1922)
Brooch. 1950s
Copper, painted enamels, twine, gilding
RFF, inventory No. 416

184. Alexander Alexeyevich Khaunov (b. 1945)
Lydia Nikolayevna Matakova (b. 1937)
Evening set. 1993
Copper, painted enamels, twine, silver plating, oxidizing
RFF

185. Tatiana Sergeyevna Mikhailenko (b. 1944)
Lydia Nikolayevna Matakova (b. 1937)
Kokoshnik necklace. 1986
Copper, painted enamels, twine, filigree balls, silver plating, oxidizing
RFF

186. Tatiana Sergeyevna Mikhailenko (b. 1944)
Valentina Vasilievna Soldatova (b. 1927)
August box. 1979
Copper, painted enamels, twine, gilding
7.5x5.1x2
RFF, inventory No. 116

187. Elena Sergeyevna Kotova (b. 1948)
Valentina Vasilievna Soldatova (b. 1927)
Brook bracelet. 1981
Copper, painted enamels, twine, silver plating, oxidizing
RFF, inventory No. 794

188. Natalia Vladimirovna Serova (b. 1954)
Nikolai Nikolayevich Mishin (b. 1955), jeweler
Expectation flask. 1989
Copper, painted enamels, twine, silver plating, oxidizing
4.5x6.5
RFF

189. Natalia Vladimirovna Serova (b. 1954)
Valery Sergeyevich Malenkin (b. 1958)
Brooch with a Still Life. 1989
Copper, painted enamels, twine, silver plating, oxidizing
RFF

190. Larissa Dolgatovna Samonova (b. 1951)
Mikhail Alexandrovich Firulin (b. 1956)
Triumph bracelet. 1987
Copper, painted enamels, twine, silver plating, oxidizing
RFF

191. Larissa Dolgatovna Samonova (b. 1951)
Alexander Viktorovich Mukharev (b. 1952)
Lilac set. 1993
Copper, painted enamels, twine, silver plating, oxidizing
RFF

192. Vladimir Pavlovich Grudinin (b. 1954)
Lydia Nikolayevna Matakova (b. 1937)
Forest Glade box. 1989
Copper, painted enamels, twine, silver plating, oxidizing
4.7x4.7x3
RFF

193. Vladimir Pavlovich Grudinin (b. 1954)
Lydia Nikolayevna Matakova (b. 1937)
Rostov Melodies set. 1989
Copper, painted enamels, twine, silver plating, oxidizing
RFF, inventory No. 53

194. Nikolai Alexandrovich Kulandin (b. 1930)
Prince Vasilko panel. 1962
Copper, painted enamels, metal frame
14.6x20.4x1.5
RFF

195. Nikolai Alexandrovich Kulandin (b. 1930)
Battle. Alexander Nevsky panel. 1978
Copper, painted enamels, metal frame
8.5x10.5
RFF, inventory No. 58

196. Alexander Vasilievich Tikhov (b. 1948)
Valentina Vasilievna Soldatova (b. 1927)
In the Meadow. 1977
Copper, painted enamels, twine, silver plating, oxidizing
7x5.5x1.5
RFF

197. Boris Mikhailovich Mikhailenko (b. 1941)
Morozko panel. 1976
Copper, painted enamels, frame — cloisonne enamel
17.3x9
Received from the author in 1976
RYMPAA, inventory No. F-20

198. Nikolai Alexandrovich Kulandin (b. 1930)
Lydia Nikolayevna Matakova (b. 1937)
September box. 1993
Copper, painted enamels, twine, filigree balls, silver plating, oxidizing
5.9x4.3
RFF

199. Alexander Vasilievich Tikhov (b. 1948)
Lydia Nikolayevna Matakova (b. 1937)
Winter Motifs box. 1993
Copper, painted enamels, twine, silver plating, oxidizing
RFF

200. Valery Dmitrievich Kochkin (b. 1957)
Alexander Kirillovich Toporov (b. 1951)
Rostov Fairy-tale casket. 1982
Copper, painted enamels, twine, filigree balls, silver plating, oxidizing
9x13x3
Transferred from A. K. Toporov in 1983
RYMPAA, inventory No. F-1644

201. Alexander Alexeyevich Khaunov (b. 1945)
Alexander Kirillovich Toporov (b. 1951)
Rostov. Stables box. 1987
Copper, painted enamels, twine, German silver
6.8x4.4x2.7
Transferred from A. K. Toporov in 1987
RYMPAA, inventory No. F-1953

202. Vladimir Pavlovich Grudinin (b. 1954)
Alexander Alexandrovich Vlasichev (b. 1956),
Bidding Farewell folder. 1982
Copper, painted enamels, twine, filigree balls, silver plating, oxidizing
7.3x8.5
RYMPAA, inventory No. F-1841

203. Valery Dmitrievich Kochkin (b. 1957)
Alexander Sergeyevich Serov (b. 1954)
Linden Bark Basket box. 1987
Copper, painted enamels, plaiting, twine
5.5x5.5x3.7
RFF

204. Larissa Dolgatovna Samonova (b. 1951)
Mikhail Alexandrovich Firulin (b. 1956)
Irina set (pendant and ear-rings). 1992
Copper, painted enamels, twine, filigree balls, silver plating, oxidizing
RFF

205. Valery Dmitrievich Kochkin (b. 1957)
Sergei Anatolievich Lebedev (1947-1993)
Autumn Melody box. 1989
Copper, painted enamels, twine, silver plating, oxidizing
4.9x4.9x2.7
RYMPAA, inventory No. F-2321

206. Valery Dmitrievich Kochkin (b. 1957)
Alexander Sergeyevich Serov (b. 1954)
Soft Snow brooch. 1989
Copper, painted enamels, twine, silver plating, oxidizing
RYMPAA, inventory No. F-2309

207. Natalia Vladimirovna Serova (b. 1954)
Alexander Sergeyevich Serov (b. 1954)
Morning Bouquet box. 1984
Copper, painted enamels, twine, silver plating, oxidizing
4x4x4
RFF

208. Svetlana Yurievna Pavlova (b. 1963), enameller
Alexander Viktorovich Mukharev (b. 1952),
Enchantment box. 1993
Copper, painted enamels, twine, filigree balls, silver plating, oxidizing
5.4x4.2x2.8
RFF

209. Larissa Dolgatovna Samonova (b. 1951)
Valery Mikhailovich Kuznetsov (b. 1961), jeweler
Rowan-tree set (necklace, bracelet, ear-rings and ring). 1993
Copper, painted enamels, twine, filigree balls, silver plating, oxidizing
RFF

210. Larissa Dolgatovna Samonova (b. 1951)
Sergei Dolgatovich Sharabudinov (b. 1956), jeweler
Miss Russia-1993 Crown. 1993
Copper, painted enamels, twine, filigree balls, silver plating, oxidizing
RFF

Contents